HIDDEN WONDERS

Library of Congress Control Number: 2020934592

ISBN: 978-0-262-53989-0

10 9 8 7 6 5 4 3 2 1

Étienne Guyon

José Bico • Étienne Reyssat • Benoît Roman

HIDDEN WONDERS

THE SUBTLE DIALOGUE
BETWEEN PHYSICS AND ELEGANCE

Translated from the French by Patsy Baudoin

Illustrations by Naïs Coq

The MIT Press
Cambridge, Massachusetts
London, England

Preface

When admiring the structure of the Eiffel Tower, the patterns that clay makes when it cracks, a bird's nest, the wavy edge of a leaf of lettuce, or even the way pole vaulters' poles bend, as if they were about to break just as they propel vaulters into the air, you may have intuited that behind the objects that populate our daily life there's strange beauty, a beauty that seems to be the fruit of a natural or willed order or underlying organization, but also of a function whose meaning we actually perceive without always being able to grasp it.

The purpose of this richly illustrated book is to reveal this beauty by relearning to see the world around us. It is the work of a team of researchers at the ESPCI (École Supérieure de Physique et de Chimie Industrielles de la Ville de Paris), and whose central themes lie where the physics of materials and mechanics intersect.

From the vaults of Azay-le-Rideau to soap foams, to a crumpled paper ball and a creeping vine bridge, the thirty-five themes addressed in this book might evoke an eclectic poetic inventory. This is not the case, of course: not satisfied to embody a fascinating dialogue among forms, forces, and functions, these subjects have in common the elegance of the shapes they exhibit, whether natural or human-made constructions.

You might rightly be surprised that the notion of *elegance* should be advanced in a book written by physicists. The term immediately evokes *haute couture*, which we will encounter in this book, but also harmony, delicacy, and simplicity. That's why elegance makes sense in many other areas too: mathematicians, for example, are happy to talk about an elegant proof that makes it possible to reach a desired result: "It was obvious; but we still had to come up with it!" By revealing hidden elegance in various fields, we also want to shed a different light on our daily lives. Our ambition is to propose a certain way of looking, an art of observing at various scales, from the mite so dear to Pascal and La Fontaine to entire civil engineering constructions.

The starting point of this project was a large exhibition coordinated around the theme of *Breaking/Flowing* (*Casser/Couler*, later at Paris's science museum, Palais de la Découverte, under the title *Ruptures*). In keeping with our frequent participation in science fairs and other public events that present science as a hands-on pursuit, we have concluded each chapter with a short, illustrated, and easy-to-reproduce experiment.

Curiosity pushed us beyond this exhibition to explore other areas. Each chapter often resulted from sometimes unusual or random encounters with interlocutors from different spheres: master folders, glassblowers, wood carvers, cellists, artists and artisans, as well as research colleagues. The diversity of this book owes them a great deal.

The blog *Hidden Wonders* at https://blog.espci.fr/hiddenwonders/ supplements this book, with videos of experiments, relevant links, and bibliographic references, not to mention our new discoveries.

We invite you to explore with us some of the hidden wonders in our everyday life!

I
BULDERS

This incredible natural lace, about 10 centimeters long and 3 centimeters wide, is the Venus flower basket's skeleton, thus named because it constitutes a natural shelter for tiny crustaceans. The lattice structure is a twisted skein of fibers, themselves composed of silica grains welded together. Thanks to their hierarchical construction, these anonymous underwater beings seem to challenge masters of architecture such as Filippo Brunelleschi, Gustave Eiffel, and Frei Otto.

THE ELEGANCE
OF SMALL AND
SLENDER THINGS

Plant, animal, and human constructions are subject to the same inescapable law: the taller they are, the more their shape must be gathered. Can a giant remain graceful?

The Cathedral of Santa Maria del Fiore in Florence, also known as the Florence Cathedral, one of the largest in Europe, is breathtakingly beautiful. Its ribbed vaults as well as its dome present shapes so simple and slender that at first we can imagine that building them was easy. Yet nothing could be further from the truth. The erection of this building began in 1300 and lasted more than one hundred years. During this period, the problem of what type of roof would cover the central part, the famous dome by Filippo Brunelleschi (1377–1446) and its impressive cupola measuring 45 meters in diameter, was left in abeyance. A real challenge for engineers—the construction of this dome alone would require forty years of additional work.

1 | *The Florence Cathedral: more than one hundred and forty years of work for a building whose capping phase was a cupola measuring 45 meters in diameter.*

At the time, vaults were traditionally built on a centring in the shape of the structure in the making [ELEGANT STONE ARCHES]. Nonetheless, although this solution might work for structures a few meters high, it was unrealistic for this colossal cathedral. How to do this, therefore? The dome finally took shape thanks to the inventiveness of Brunelleschi, an artist and builder of genius: he imagined an ingenious, self-supporting framework, consisting of superimposed bricks canted layer by layer. To complete the project, Brunelleschi also designed cranes and other innovative devices, as evidenced in some of Leonardo da Vinci's drawings.

Galileo Looks into Hell

But why is it so difficult to build big? To understand this, let's take a short detour via Galileo. Galileo is perhaps best known for his defense of heliocentrism, which started his troubles with the Catholic Church, but his contributions to modern science are many and varied. In particular, he had the brilliant idea of comparing objects of different sizes in order to study the relative effects of physical forces. He carried these studies throughout his life, and he gave them shape in a major work, *Two New Sciences*, which he wrote after being condemned, and whereby he invented the discipline of the strength of materials. The seeds of Galileo's reflections can be found in a lecture he gave before the Academy of Florence at the beginning of his career. That presentation dealt with the geometric form of hell in Dante's famous *Inferno*. The site was described as a conical abyss dug beneath Jerusalem and covered with a dome (the earth's crust), whose thickness Galileo was discussing. He imagined that the cover of the cave of Hell, which is a hundred thousand times wider than the dome of the Florence Cathedral, should therefore simply be a hundred thousand times thicker: a reasoning of proportionality that seems incontrovertible, but which assumes that identical shapes ensure the same solid strength.

Size Matters

Galileo's reasoning is obviously wrong; if not, the construction of the Florence Cathedral would have not posed any problems—no more, in any case, than those involved in erecting a simple chapel. Galileo himself recognized his mistake at the end of his life, when he was under house arrest in Florence. In *Two New Sciences* he wrote that if the dome covering Hell were subjected to the rule of proportionality, it would simply collapse under its own weight.

What Galileo had finally grasped was that if one multiplies the size of an object by ten, its weight, a thousand times greater, is distributed over a surface that is only a hundred times larger. To build big, then, is to constantly fight against forces of compression that increase with height, even if we take care to expand the base of the building. That is why, if one wishes to confer grace and a slender shape to a building, one must favor a modest size. It is a law of nature: *Parvus est bellus*, or "small is beautiful!"

The Size of Giants

Examined closely, the animal world offers many examples of the limitations imposed by earth's gravity. Thus, the skeleton of a shrew accounts for only 5 percent of its total weight, while that of an elephant, though relatively similar in shape, is already 20 percent of its own weight (fig. 2). Galileo himself had already acknowledged these consequences in his treatise: "Increasing the animals enormously in height can be accomplished only by employing a material that is stronger than usual or by enlarging the size of the bones, thus changing their shape until the form of the animal looks monstrous."

Large animals, despite their apparent power, do not move with agility: an elephant is unable to jump like an antelope because it would submit its skeleton to too much stress. Similarly, one can possibly imagine a dog carrying a fellow dog on its back, but a horse can't accomplish such a feat.

On the other hand, ants raise colossal loads compared to their weight, but this is not a feat!

What about marine animals? The constraints of gravity disappear for animals that float in a liquid medium whose density is very close to that of their bodies. The largest animals ever to exist on Earth are blue whales, which, despite their size, do not feel their weight. Moreover, their skeleton differs little from that of their smaller cousins such as dolphins, but for the magnifying effect (fig. 3).

The plant world is no exception to this rule. While it is growing, a tree must, respecting Galileo's rule, strengthen its trunk, not proportionally to its size, but more rapidly. In its early years, a very slender trunk will suffice, but once it is a venerable giant, it will have to become much thicker and stockier to resist the load. Indeed, a tree that is too slender tends to bend under the effect of its own weight, like a playing card whose oppo-

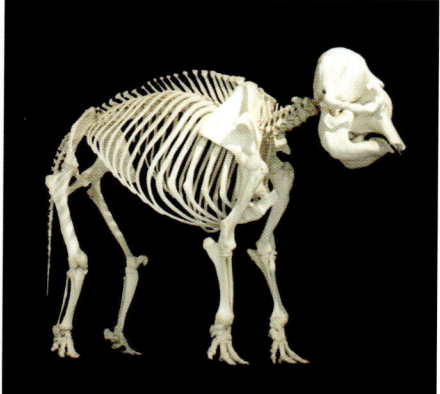

2   *The skeleton of a 10-centimeter-long mouse accounts for only 5 percent of its weight, whereas that of an elephant weighing several tons accounts for 20 percent of its weight, which results from the constraints imposed by gravity.*

site edges are brought together between thumb and index finger. Mechanical engineers speak of buckling [BEAMS BEND BUT DO NOT BREAK]. How do "simple plants" self-adapt to the effects of gravity? This is a mystery that researchers are trying to solve.

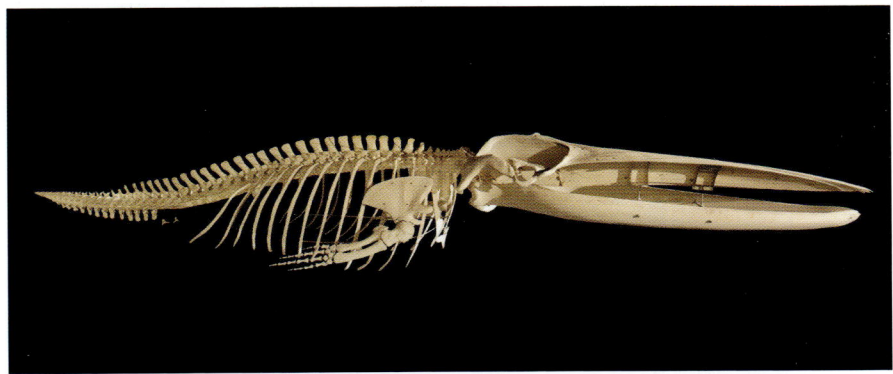

3 | *A whale (top) can weigh up to a thousand times more than a dolphin. Yet the skeleton of these two cetaceans is in the same proportion to their total weight.*

EXPERIMENT

Roll two sausage-shaped pieces of play dough of different sizes, while maintaining the same length-to-diameter ratio. Which one will remain standing? In the illustration, two columns of 5 millimeters and 1 centimeter in diameter were prepared for respective lengths of 5 centimeter and 10 centimeter lengths. These columns thus have the same shape, the size of the small column being just half the size of the big one. The bigger column, even though it is wider, can't stand upright and bends miserably under its own weight, unlike the small one. Thus, these play-dough columns can hardly reach a few dozen centimeters in height!

4 | *Columns of the same proportions: the bigger one tends to collapse under its own weight..*

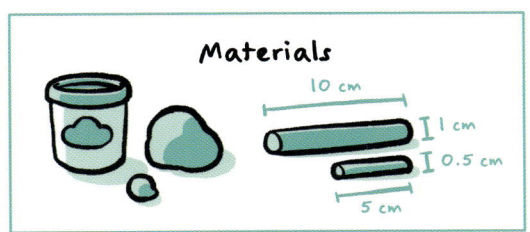

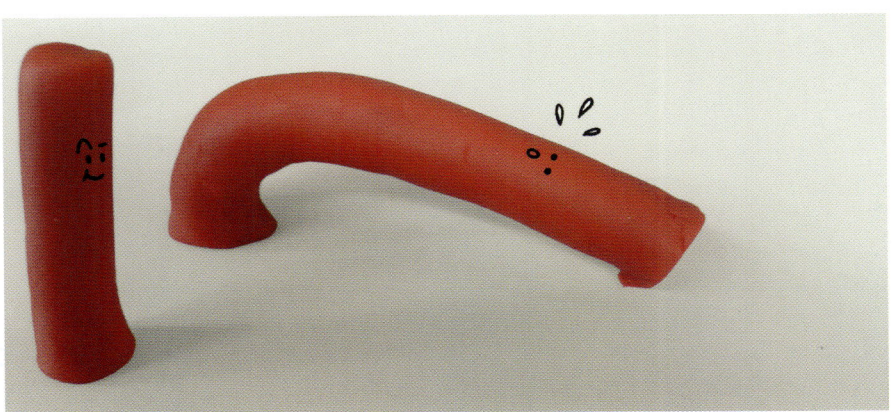

AZAY-LE-RIDEAU OR ROOFS OF BEAUTY

Each beam of a structure fulfills a well-defined mechanical role, which ensures the stability of the whole. As illustrated by the Château d'Azay-le-Rideau in the Loire Valley, a robust and durable frame remains elegant.

Climbing a ridge, I admired for the first time the Château d'Azay, a cut diamond set along the Indre, mounted on stilts wrapped in flowers.
—Honoré de Balzac, *The Lily of the Valley*

1 | *The framework of Azay-le-Rideau dominates the roof space. The oblique girders or rafters (R) are held in place by the combined tie-beam (T-B) and the king post (KP).*

To visit Azay-le-Rideau is to enter into one of the temples of the Renaissance. Built on an island during the reign of King Francis I, the chateau is famous for its sumptuous setting in waters formed by the two arms of the Indre river, which reflect its carved stone facades. But there's another architectural jewel to discover. Just take the beautiful, sky-lit staircase to the top, directly under the dome. What a dazzling masterpiece! This structure feels strong and light—a combination found in the elegant framing of old barns. But what are the principles honored in this construction?

From Rafters to King Posts

Completed in 1522 and built like a cathedral, the architecture of Azay-le-Rideau includes a high ceiling that echoes the steeply pitched roofs covered in black slate. The elements that prefigure the shape of the cathedral's roof are oblique beams. Forming an inverted V, they meet at the top and support the roof. The feet of these *rafters* would be able to deviate if they were not connected by a 12-meter-long horizontal beam, a *tie-beam*, carved of a single oak trunk. This bracing element prevents the opening of the V-shape formed by the rafters and the application of excessive horizontal forces on the walls that support the frame.

The last centerpiece of the structure is the *king post*, a vertical mast that joins the middle of the tie-beam and the ridge beam. Contrary to appearances, this feature is not compressed; its role is the opposite of suspending the tie-beam to the upper part of the frame. Without this relief, the horizontal tie-beam with its long reach could eventually bend under its own weight. The upper frame is also reinforced by secondary elements: they are there to limit the bending of the rafters.

Folding without Breaking

The oaks providing the timber of the Azay-le-Rideau frame were felled some six hundred years ago. The beams are so resistant to breakage because the fibers that constitute the wood are oriented length-wise. If a crack appears under the impact of significant stress, then the discontinuity among the fibers will limit propagation of the crack through that section. In a way, the wood adopts the same strategy observed in the creation of mother-of-pearl [SHELLS AND MILLE-FEUILLES].

While wood is an ideal candidate for horizontal beams, stone seems much less suitable. Under its weight, this beam would flex as its center was pushed down, which would strain its lower part [ELEGANT STONE ARCHES]. Now, although stone, like concrete, easily supports the compression induced by the load of a pillar, it offers only poor resistance to traction. A crack would appear very quickly on the underside of the beam and spread catastrophically through its section. Enough about the stones!

The Art of Breaking Sticks

The risk that its horizontal beams might change shape has therefore partly guided the design of the frame. But why is this risk so great? Given the imposing section of these beams, one might imagine that they would have no trouble withstanding their own weight. The architect of Azay-le-Rideau's frame conceived of his plans on the basis of empirical findings. In 1637, a century after the construction of the chateau, Galileo addressed the same issues in his *Two New Sciences*, when he wondered why it is easier to break a stick on your knee than to pull it along its axis [THE ELEGANCE OF SMALL AND SLENDER THINGS]. He suggested that the answer should be sought in a simple lever effect (fig. 2). Indeed, the transverse forces are multiplied by the lever arm, in this case the length of the beam. As an order of magnitude, this rate of amplification of the forces is easy to calculate: it is the ratio of the length of the beam over its thickness. This is why elongated objects are easy to bend and tend to deform by flexion.

An Upside-Down Ship

Curiously, the architecture of Azay-le-Rideau's vaulted ceiling is reminiscent of the hull of an upside-down boat. This is not a coincidence: it is built according to the same principles. The weight of the frame is replaced by the pressure exerted by the water on the hull. The hull of a ship, however, is subjected to more complex loads than a roof, which must essentially withstand only its own weight. The ship also undergoes torsion

2 *This drawing by Galileo illustrates the fact that a horizontal beam does not easily withstand the bending imposed by a load E placed at the end of the beam.*

stress when, for example, it is struck by waves on its side, which makes it take in water through the cracks of the hull. It was not until the nineteenth century that wooden hulls benefited from additional, diagonal beams to counterbalance this torsion. Did you know that the fishermen at Équihen-Plage near Calais on the English Channel used overturned hulls as roofing for their homes? The region was also called the "land of upside down keels."

3   *From ancient times until the present day the inhabitants of Équihen-Plage turned over the hulls of their boats to serve as roofs for their single-story houses.*

EXPERIMENT

How do you increase the flexural strength of a sheet of paper? By folding it like a fan!

The thinner a sheet of paper, the more flexible it is. Placed flat between two books (1), it will immediately bend under its own weight (2). A trick, however, makes it possible to considerably increase its resistance to the point that it can support the weight of a cup: fold it like a fan!

Not convinced? Fold a standard sheet of paper every centimeter. This has the effect of amplifying its apparent thickness: since the sheet is one-tenth of a millimeter thick and the fold's amplitude is 1 centimeter, the gain in stiffness is colossal—on the order of 5,000-fold. That's an ingenious strategy to get materials that are both robust and light!

Cardboard boxes use the same principle. Looking at the cross section of a piece of cardboard, you will observe the same undulated structure. Many variants of this pattern exist in nature and in engineering, which aim at increasing the apparent cross section of objects that have to withstand bending: hollow stems, the I-beams of modern buildings, and so on. However, the use of such light structures can be risky: they sometimes give way catastrophically [BEAMS BEND BUT DO NOT BREAK], in contrast to solid rods, which bend gradually under a load.

Materials

1.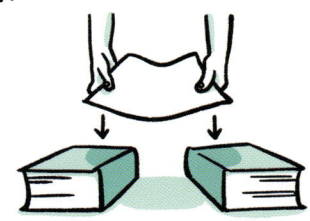

2.

PLOP!

3.

fold it

fold it

4.

THE EIFFEL TOWER

The Eiffel Tower, icon of France's capital, is at the heart of an industrial revolution—that of metal constructions. Its shape continues to be intriguing, but did you know that it embodies an original, technical feat by Gustave Eiffel?

A Metal "Carcass" a Thousand Feet Tall

What is more emblematic than the slender, curved shape of the four sides of the Eiffel Tower? Erected to rise above the Universal Exhibition of 1889, this tower initially had no other goal than to reach beyond the mythical height of 1,000 feet, thus displaying the boldness of engineers and the triumph of the Industrial Revolution. Before it was built, critics fumed over it, with Guy de Maupassant describing it as "this tall and thin pyramid of iron ladders, an awkward and giant skeleton, whose base seems made to bear an imposingly massive monument to Cyclops and aborts into a ridiculous and thin factory chimney profile."

> 1 | *Thanks to Gustave Eiffel's genius, this "awkward and giant skeleton" has become the symbol of Paris. Its intriguing architecture is in fact designed to resist wind in the most effective manner possible.*

To this Gustave Eiffel replied that "the tower will have its own beauty"; that the conditions of its design compelled to resist forces "conform to the secret conditions of harmony," nothing less; and that "the first principle of architectural aesthetics is that the essential lines of a monument are determined by a perfect appropriateness to its location." The Eiffel Tower would have the elegance of an optimal shape, in a sense designed for and by the forces it would have to resist. As a result, its abstract form, its very contrast with the classical Parisian landscape, even beyond the various functions that it has experienced throughout its history, amount to the Eiffel Tower no longer being questioned by anyone.

A Wind Tunnel at the Champ-de-Mars

What led Gustave Eiffel to create such a distinctive shape? Before metal was used for construction, it was a building's weight that determined its maximum height. The tallest building erected before Eiffel's feat was the Washington Monument, an obelisk, 169 meters tall, and completed in 1850—not without some difficulty since the weight of the granite would cause the soil to compress and complicate construction. The Washington Monument was completed only a few years before the Eiffel Tower. This obelisk, however, was not much taller than the Cologne Cathedral (at 156 meters) built six centuries earlier. Thanks to iron, it became possible to reach much higher and more lightly, especially since this steel Meccano-like structure was easier to assemble. The Eiffel Tower would reach the mythical 1,000-foot-high bar, more than 300 meters, thanks to its 18,000 metal pieces assembled using more than 2.5 million rivets. The nineteenth century was a new Iron Age; thanks to iron's modest yet indispensable partner, it was also "the Age of the Rivet."

At these heights, winds are the main enemy of such works. Gustave Eiffel had already discovered this earlier, particularly when building the pillars of the Garabit viaduct in the Cantal region of south-central France. He would eventually satisfy his lifelong passion for aerodynamic forces by dropping objects from the tower and subjecting them to the wind

2   *Before the Eiffel Tower, the Garabit viaduct already had pylons that narrowed at their tops. The wind's effects are mainly exerted at the level of the deck. The pointed-pillar shapes are calculated to eliminate the flexion of individual beams.*

created by their fall. He would then invent the principle of the wind tunnel—the Eiffel wind tunnel—which is still in operation today. First installed at the foot of the Eiffel Tower in the public park of the Champ-de-Mars in 1909, it would be moved a few years later to a suburb of Paris for tests in the nascent field of aviation.

Form and Function

What, then, is this shape that best resists the wind, the form that makes it possible to achieve the highest structure for a given quantity of material? It is known that to ensure the strength of elongated structures, forces must be borne along the axes rather than transversely. To break a piece of wood, for example, you don't pull on it along its axis, but rather you take it to your knee and pull on both ends. In other words, you apply forces perpendicular to its axis, which is much more effective.

In the case of stone structures limited by their weight, it is therefore the shape of a column or an obelisk that bears the greatest loads along its axis and is therefore the most suitable for building high. In fact, it is this sleek shape that one finds in many towering masonry projects.

But wind pushes laterally, and you have to adopt another strategy. Gustave Eiffel then devised a framework in which at each level the four ridges of the tower play the role of flying buttresses by gradually spreading out nearer their base. This new shape is not rectilinear, as were all tower projects proposed until then; it tapers all the way to the top. In 1884 Eiffel and two of his engineers filed a patent "for a new arrangement to build piers and metal pylons that can exceed a height of 300 meters."

3 | *An initial 1,000-foot high "pylon" project already contains the idea of emptying the central part of the tower. On the right is the lineup of monuments that, according to its detractors, the Eiffel Tower overwhelms. Note, too, the delay between when this plan was drawn in June 1884 and the end of construction less than five years later. Who can do better today?*

EXPERIMENT

To understand the shape of a tower, we can begin as Gustave Eiffel did by reasoning about the piles that support the Garabit viaduct, which was built before the Eiffel Tower. These piles have to resist the horizontal force of the wind, which is mainly exerted on the upper deck. If we leave aside the lesser impact of the wind on the vertical piles, we realize that the triangular shape of these piles optimizes their mechanical resistance to the wind, as the following experiment shows.

Cut a rectangle out of a piece of Bristol board and draw folds along two parallel lines to create three equal parts. By folding back and then stabilizing the third edge using adhesive tape (1), you'll create a triangular shape to affix to the table (once again with adhesive tape). That's the viaduct's pile. Push horizontally on the top of the triangle to simulate the pressure of the wind on the deck (2). You'll find that the building is quite stiff (as long as you don't push too hard!): the force is transmitted in traction on the side of your finger and in compression on the other side, without any unwanted bending [AZAY-LE-RIDEAU OR ROOFS OF BEAUTY]. On the other hand, if you press halfway up one side of the triangular piece of cardboard, you'll feel far less resistance because the side will bend.

As for the Eiffel Tower, it has to resist a wind force distributed along the pillars. How can the tower avoid bending effects and achieve an optimal shape? Eiffel draws the four curved beams that are the edges of the tower so that they play the uninterrupted role of crutches. On each floor, these beams have to be oriented exactly towards the point where the wind forces come to exert their load on the upper part, thus forming an optimal triangle. Starting from the top, and repeating the procedure all along the tower, it adopts this flared—and so elegant—shape towards its base.

Bristol paper

Materials

1. The bridge's pile

~4cm

2. Wind

INNER
BALANCE

Imagine it raining long needles that suddenly freeze forever, as if suspended in air: artist Kenneth Snelson's Needle Tower *seems to defy the laws of nature. What makes a work like this possible? It's thanks to tight cables and a clever balance between the tension and compression of the rods.*

At a stone's throw from the White House, the Hirshhorn Museum of contemporary art is itself a full-fledged work with a beautiful sculpture garden. There stands an incredible tower, 30 meters high, made of needles that seem to be floating in the air, like six sticks from a bunch of pickup sticks suddenly reaching for the sky.

Seen from its interior, the structure of the *Needle Tower* outlines six-pointed stars, but its artist, Kenneth Snelson (1927–2016), insisted he had not injected any sort of symbolism. Above all, he saw himself as the architect of a unique construction that was unprecedented until the beginning of the twentieth century.

> 1 | The Needle Tower *erected by Kenneth Snelson in 1968 in a sculpture garden in Washington, DC. How are the rods, held together by cables, suspended?*

How is such a mechanical feat possible? Its secret lies in the thin, stretched cables that connect the rods to each other. By ensuring a delicate balance of internal forces, they guarantee the stability of this at once light and elegant sculpture.

When viewed from afar, Washington, DC's surprising *Needle Tower* looks like an Eiffel Tower. But send the two into satellite flight, and that's where their resemblance stops: in the absence of any gravity, Gustave Eiffel's structures would not be subjected to any internal constraint, unlike Kenneth Snelson's work.

Born with Aviation

To identify the historical ancestors of such structures, you have to turn to the early years of aviation, when the idea of coupling rods and cables emerged: think of the wings of the first biplanes.

2   *In Brisbane, Australia, the Kurilpa Bridge was partly built respecting tensegrity codes entailing use of tight cables to connect rods. Spanning 470 meters, this is the largest structure of this type in the world.*

Works of contemporary architecture subsequently used this same combination of cables and rods; but examples of large-scale use, such as the Kurilpa Bridge in Brisbane (fig. 2), remain marginal. Such bristly structures are indeed difficult to cover and, unless extremely taut, they tend to bend out of shape under the effects of wind. Temporary, deployable, and very light shelters are built in a similar, though less monumental way. Instead of using cables to connect the rods, designers prefer taut membranes (fig. 3). Their structure hearkens back to textile architectures that remind us of the shape of soap films [SURFACE TENSION].

Tensegrity: A Neologism

American architect Richard Buckminster Fuller (1895–1983) summed up the complexity of such balances by calling this type of structure an archipelago of "islands of compression in a sea of tautness." Buckminster Fuller's name is associated with the neologism "tensegrity," a portmanteau of "tension and integrity." He is best known for his geodesic dome in Montreal.

The "skwish," a toy described on page 32, illustrates the notion of tensegrity, although for specialists this example does not cover the many structures balanced in this manner, which have existed much longer.

The Strength of Bike Wheels

Even if the downsides of this kind of construction have restricted its use in architecture, history shows that humans have from time immemorial designed objects or buildings within which the internal stresses of tension and compression are balanced. Such is the case of the spoke wheel, which was invented about two thousand years before our era in the Eurasian steppe and made it possible to replace solid and studded wheels. Relatively thick wooden spokes are compressed by the rim, which ensures the mechanical strength of the wheel. Bicycle wheels as we know them operate on the opposite principle: thin and light spokes are subjected to high tensile force (that's what a spoke wrench applies), which puts the rim into compression. Mostly, we owe this ingenious solution to French inventor Eugene Meyer, a pioneer in the development of the first bicycle, known as a "penny-farthing."

Another example of the balance of forces are cable-stayed bridges, like the Millau Viaduct in the Aveyron region of southern France. The *stay* hearkens to the stretched cables that help a mast withstand the wind that swells the sails. In the case of the bridge, the stays are used to hold up the deck and are placed symmetrically on either side of its pillars, to balance the horizontal component. The bridge is thus internally balanced and does not need the strong anchoring of cables at the ends, which is required on a suspension bridge such as the Golden Gate Bridge in San Francisco [NECKLACES AND CATENARIES].

In Living Structures Too

Nowadays, the balance of internal constraints is also invoked to describe certain architectures of life. In the plant world, the superficial layers under tension compress the heart of the plant. If you peel off one side of a rhubarb stalk or slash the end of a radish, you will notice that the stalk or radish changes shape spontaneously owing to the loosening of internal tensions.

In vertebrates, the mechanical behavior of disjointed elements and bones (which ensures the flexibility of the whole) is made possible thanks to a network of tendons. This subtle compensation of opposing forces is essential for the mobility of living beings: a dancer standing on toe manages a delicate balance between the compressed bones of her leg and the tension in her muscles. This enables her mobility, which a frame such as a cast set on a broken leg does not.

On a smaller scale, tensegrity also comes into play in the mechanical behavior of cellular organisms. In the cytoskeleton of eukaryotes, a complex structure of filaments made of polymers and other more rigid constituents such as microtubules serves in a way to reinforce cellular forms. One must note, however, that this simplified vision does not do justice to the diversity of biochemical behavior (such as the effects of osmotic pressure on the walls) in these natural constructions!

3 | *A prototype of a deployable structure developed at Tokyo Science University. Weighing a total of 600 kilograms, it can cover a floor area of 150 square meters. The rigid disjointed rods are inserted into slides arranged in the polyester fabric.*

EXPERIMENT

The "skwish" is a toddler's toy with small wooden dumbbells that form a structure held by elastic cables and whose shape can be changed. Making a skwish yourself can help you appreciate the subtle balance between forces of compression and forces of tension.

Use six wooden sticks that are slit at their ends and six rubber bands the same length as the sticks (dimensions can obviously be bigger or smaller).

Sliding the rubber band in the slits, start by wrapping a rubber band around each stick. Then create an H-shape by inserting the ends of a pair of parallel sticks through the middle of the rubber bands of both other sticks (1). Insert the two remaining sticks (pair 3) through the middle of the rubber bands of the initial pair of sticks (2).

The end requires a bit more care (using four hands can help!). It involves inserting the remaining rubber bands into the two corresponding slots. By pushing a bit, you will manage to create a beautiful, regular volume with twelve vertices—an icosahedron.

Materials

Basic operation

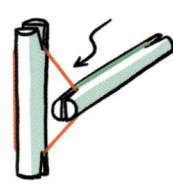

1. H-shape

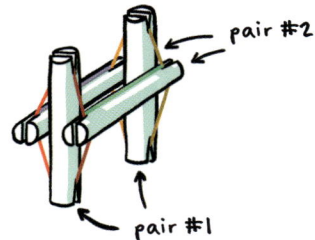

pair #2

pair #1

2.

pair #3

3.

SURFACE
TENSION

The Olympic Stadium in Munich is a fine example of architectural membranes. As incongruous as it sounds, to conceive of it architect Frei Otto was inspired by soap films stretched on a frame.

Have you had the opportunity to check out the latest designs for camping tents? We know that habitats made of skins or canvas stretched on a frame or under a central pole have been universally used by nomadic peoples, saving weight and easing handling for frequent moves. But, nowadays, tents have become technological objects; some can even unfold automatically. In contrast to older ridge tents, these new ones are held up by strongly arched frames and resemble igloos. The fabric's tension, generated by the incorporated beams' flex, makes it possible to eliminate the folds.

| 1 | *The buildings of Munich's Olympic Stadium evoke giant soap films stretched over a nonplanar frame.* |

Inspiration in Soap Films

Architects have not been blind to the beauty of the spontaneous shapes that canvas surfaces take; they have been inspired by it to imagine and build large, permanent structures. The most famous example of this is the stadium that architect Frei Otto (1925–2015) designed for the Munich Olympics in the early 1970s. Stretched at the top of vertical pillars, its cables are spread across the upper surface of the stadium. Covered with glass panels, they provide a broad, continuous cover, just as taut sails would have in their place.

The resemblance is not accidental. In fact, Otto was inspired in his stadium design by a special kind of canvas: soap films. Otto plunged into a bath of soapy water a twisted wire frame shaped for use in his

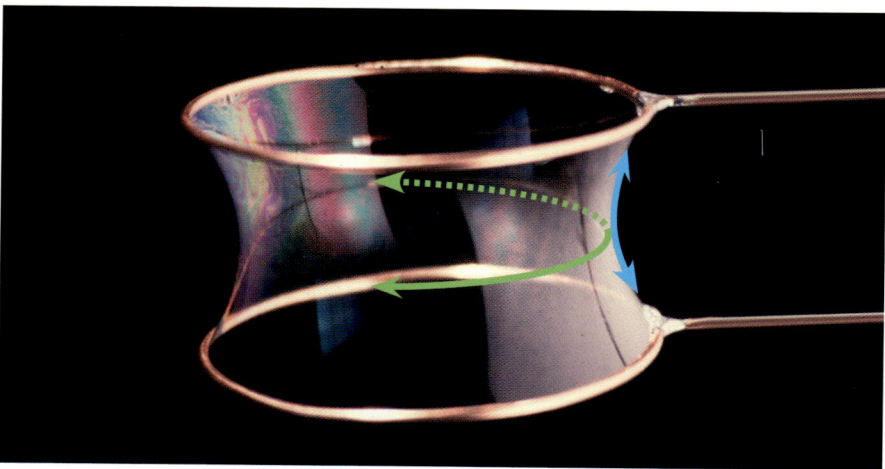

2   *Soap film stretched between two loops has what is called a catenoid shape, such that at every point the two curvatures of the surface are equal and opposite.*

Frei Otto often used soapy water films resting on a suspended frame as a model for his architectural creations, such as the elegant Dance Fountain (Tanzbrunnen) in Cologne, Germany.

architectural project. Then, at some length, he analyzed the geometry of the soap film resting on the metal.

Soap films are the equivalent of stretched canvas because of the capillary forces that pull on the films; these forces are analogous to the traction exerted by the frame on the fabric. Frei Otto's work on soapy forms would also lead him to create the Institute of Lightweight Structures in Stuttgart in 1974. Like Gaudí and his creation of the chain-shaped, vaulted ceilings of the Sagrada Família [NECKLACES AND CATENARIES], Otto drew inspiration from natural phenomena.

Stretched cloth and soap films exhibit similar characteristic shapes: these are curved at each point, both upward and downward in perpendicular directions. This is also the shape of a relief in the immediate vicinity of a mountain pass or *col*. It corresponds to a general shape that bears the evocative name of *saddle*. But why do these taut surfaces make saddle configurations appear everywhere? Cols have a geometric peculiarity (fig. 2): there are lines that curve upward (the ridge line that rises on each side of the col) and other lines that curve downward (the path taken to cross the col). Now imagine that these drawn lines are the tension lines of a stretched sail, and you'll realize that those curved upward tend to pull and bring the surface upward, as opposed to those curved downward that tend to lower it. Taut, nonplanar surfaces must therefore subscribe to a delicate balance, which only these saddle shapes can do. QED!

Mathematics and Minimal Geometries

Is it possible to go further and calculate the exact shape of the saddle? For soap films, certainly, because they represent an ideal situation. It has been shown that such film takes on a shape that corresponds to the minimum surface that can be deployed on the frame.

The physicist, astronomer, and mathematician Joseph-Louis Lagrange (1736–1813) was the first to describe the minimal surface needed to rest on a given contour. He proved that such a surface has a zero mean curve, that is, that any local curvature in one direction is compensated for by a curvature in the other direction at the same point. Yet Lagrange was only able to describe the simplest form, a *plane*. Mathematicians have since become passionate about these surfaces they call *minimal*, finding new, increasingly complex shapes, and relying today on computing power to visualize their creations (fig. 4).

4 Michael Foster's Inversion represents a minimal surface: it joins a vertical circular edge and a loose clover knot. The black border consists of a single line that makes three loops.

EXPERIMENT

With a metal wire, shape a ring extended by a rod to hold it. Immerse the ring in a glass of soapy water (1). It will come out with a flat film stretched over the edge of the wire because the surface-tension forces tend to reduce the area of the surface as much as possible (left).

To prove that such a film behaves like a sail, blow on it gently: the surface inflates while resisting against the air pressure (right), as would a tense and billowing sail. Now twist the ring: the soap film will take on the shape of a saddle, which is the minimal surface resting on the support. A whole menagerie of minimal surfaces can be obtained by varying the shape of the frame.

Let us complicate this game by using the wire to shape two parallel circular loops linked together (at the center). After plunging the frame into the water, it can be brought out with an open soap bubble. And—surprise, surprise—the tube of soap is not a cylinder. The film has an hourglass shape: it is thinner in the middle. Mathematicians have given this form, shown in figure 2, the learned name of *catenoid*.

And since we have associated bubble problems and fabric tension, why not use a nylon stocking to reproduce a shape close to this catenoid? To do this, take two balls that are barely bigger than the stocking's diameter; place them at two different heights inside the stocking (2). If you pull a little on both ends of the stocking (which initially was cylindrical), you will find that it looks like the saddle's double curvature, just like the film between the two loops!

Materials

1. Soapy mix

2. Stretched stocking

41

Pascal Oudet, *Dendrochronologie: Mailley*, 2015.

II

CREATING
SHAPES

Oak, a symbol of strength, can also be flexible. When it is cut into millimeter-thin sheets and dried, an astonishing transformation takes place: each slice turns into a sort of plate with a wavy edge, like a gigantic potato chip. What is the secret to creating such shapes? When it loses water, wood retracts more along its edges, and its perimeter shrinks. A thick block would crack, but a slice deforms according to its imperfections. More generally, the shapes and deformations of natural or human constructions account for force fields that solicit them. Such is the case of the spherical shape of soap bubbles or the elegant arc of a pearl necklace.

AND WE THINK OF BUBBLES AS FRAGILE!

Subjects of wonder at all ages, soap bubbles are shaped by the tension forces exerted on their surface.

On the occasion of accepting his Nobel Prize in Stockholm, French physicist Pierre-Gilles de Gennes (1932–2007) ended his speech on "Bubbles, Foams and Other Fragile Objects" by projecting an image of the painting *Soap Bubbles* by Jean-Simon Chardin, one of the great genre painters of the eighteenth century. Mischievously, de Gennes offered the following quatrain, daring to compare the vanity of honors with the fragility of a soap bubble—a provocation in front an audience of Prize winners!

Let's have fun. On the earth and on the waves
Unfortunate is he who makes a name!
Wealth, honors, false glory of this world,
All of it but soap bubbles.

1   Jean-Siméon Chardin, *Soap Bubbles* (1734).

Among the countless topics he has covered over the course of his career, Pierre-Gilles de Gennes chose to devote his speech to the physics of bubbles. In the painting, the curious youngster in the background looks on stunned, caught up in the suspense: will the bubble burst and project myriad droplets, or will it detach and go on a brief trip in the air? An undeniable power to fascinate emanates from these "fragile objects." Such as, why are bubbles round?

Bubbles, Drops, and a Rainbow

It is forces acting on the surface of bubbles that explains their spherical shape. From a geometric point of view, they force a bubble to take the shape of the minimal area needed to contain the volume of air that is trapped when blown into it. And a sphere is what satisfies this constraint. For a similar reason, drops of water suspended in the sky are also round: do you know that it would be enough for these drops to deviate from a sphere a few percentage points to alter the rainbows that come with wet skies? Indeed, a rainbow is created by a set of multiple reflections in each drop, which depends very acutely on the exact sphericity of these liquid beads.

Bubbles lose their balloon shape as soon as they come up against a solid. They then adopt a semi-spherical shape, connecting perpendicularly to the surface. In contrast, while retaining a spherical, cap-like shape, a drop of water more or less spreads out on a plane, depending on the liquid's chemical affinities with the surface.

Tension in the Air

But let's not lose track of the bubble. Let's look at its birth, then its growth, and its final shape. Imagine you've attached a tiny drop to the end of a straw, like a glassblower who hangs a drop of molten glass at the end of a hollow blowpipe [STATES OF GLASS]. By blowing continuously, you inflate the inside of the bubble that is taking shape. The internal overpressure is higher at the beginning of the operation, when the curvature of the bubble, that is to say the inverse of its radius, is high. Mathematician Pierre-Simon de Laplace tells us more precisely that this overpressure is proportional to the curvature of the bubble, but also to a physical quantity that is specific to the interface of the material, namely its *surface tension*.

Surface tension is required to increase the surface of the bubble. In a way, it plays the same role as the elasticity of an inflated balloon: you've probably already noticed that the pressure needed to inflate a balloon is strongest when you start to inflate it. It is also surface tension that is responsible for our wet hair sticking together [WET HAIR] and for the grains of a sandcastle cohering [THE SECRET OF SANDCASTLES].

2 For a physicist, the cyclone formed on a soap bubble is a miniature model of its atmospheric counterpart. At the scale of Earth's radius, the atmosphere amounts to a thin layer. The colors come from the light interferences in the film and reveal the variations in thickness.

Piercing a Bubble

The fragility of a bubble is relative. This thin shell of water is covered on both sides by two extremely thin continuous walls made of side-by-side soap molecules. The thickness of the film is very low, from a few micrometers to a few nanometers in the case of the thinnest. Despite its thinness, a bubble is robust. It is even possible to pierce it without it bursting by using a needle wet with soapy water!

When the thickness of the film is of the order of size of the wavelength of light, bright and sumptuous iridescences appear thanks to the interplay of interferences among the light waves that are reflected in the thickness of the film. The color changes thus tell us about thickness variations, owing, for example, to flows in the film (fig. 2). This is why a simple modulation of the hues is enough to visualize the wake induced by the displacement of an object that crosses a bubble without piercing it.

Do you know that it is even possible to produce real miniature tornadoes in a half-bubble of soap placed on a slightly heated plate? At the base of the film, the water heated by the plate begins to rise like a hot-air balloon. In contrast, the liquid at the top of the bubble cools and tends to fall back under its own weight. These movements of opposing fluids produce magnificent whirlpools.

Finally, if you're patient, you can observe the continuous unraveling of colored fringes on a bubble carefully protected from any drafts. These successive fringes attest to the film's decreasing thickness as its water drains. You'll eventually even see an area so thin that light does not reflect on it. Newton spoke of *black film* to describe this liquid layer of a few nanometers that is reminiscent of a hole. But think again: although invisible, the film, which now consists of only two soap walls facing each other, is still there. If you put your finger through this so-called hole, the bubble bursts instantly.

To Die in a Rain of Droplets

Recent developments in rapid imaging have given new impetus to the study of the bursting dynamics of soap films (fig. 3). Observed in slow motion, the few thousandths of a second it takes a bubble to die have revealed physical mechanisms of unprecedented richness and ineffable beauty. The bursting process takes place by creating a hole that quickly invades the entire bubble. During this process, the exploded part forms a mist of droplets, while the rest of the suspended bubble still outlines a portion of the sphere.

3 — *A pierced bubble shrinks under the effect of surface tension. It leaves liquid filaments that rapidly break up into microdroplets of water.*

EXPERIMENT

It is easy to make a flat film of soap: take two thin strips of plastic or wood and connect them with two cotton threads attached to their ends. Dip this frame in a dish of soapy water, making sure that the stems and wires are wet and close to each other (1). Then take this frame out of the water by carefully separating the sticks (2) and (3): you have just created a liquid film. You can, with care, use the frame as a racket and bounce bubbles on it!

After this bubble-pong game, by placing the film upright, you should see some pretty colorful interference patterns (below) that evolve as the film thins under the effect of gravity.

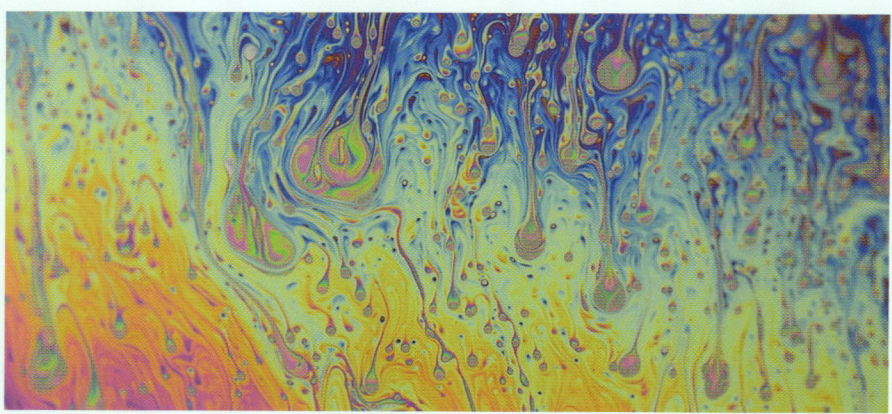

Materials

1.
2.
3.

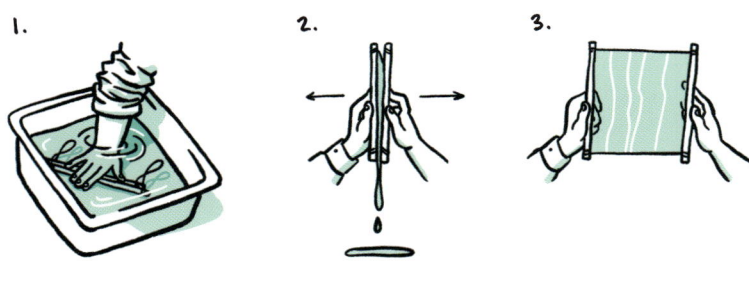

Soap-pong

THE TRAGEDY
OF FOAM

The law of the jungle even operates among foams: the largest bubbles eat up the smaller ones. A struggle in vain for survival, since all bubbles inevitably end up bursting.

If you've taken the time to observe the dance of bubbles in a glass of champagne, you may have noticed that they are born of defects or impurities present on the walls of the glass, and they rise toward the surface in beautiful single files. Why? It's the carbon dioxide produced by the fermenting yeast, which is present in this precious liquid and inflates each of these strange bubbles that rise like hot-air balloons. Having reached the open surface, some bubbles burst immediately, with a familiar fizzing sound, while spreading aromas; other, more resistant bubbles join their neighbors and form an elegant foam that reflects the quality of champagne.

1 | *At the end of their ascent, the bubbles gather in a thin sparkling foam on the surface of the champagne that adds to the pleasure of tasting it. The physical characteristics of the drink, as well as the bursting bubbles and the accompanying fizzing sound, are currently the subject of exciting research.*

But what is *foam*? Physicists describe *wet foam* as the state of bubbles that have gathered but do not yet adhere to each other. Over time the water between the bubbles is eliminated under the effect of gravity, and the walls of the bubbles merge: they then form a *dry foam*—a deceptive term because their walls are liquid films on the order of one-thousandth of a millimeter thick.

As with the soap bubbles [AND WE THINK OF BUBBLES AS FRAGILE!] these films of water are sandwiched and stabilized by two thin layers of *surface-active* molecules. This adjective denotes elongated, nanometric molecules, whose one end "likes" contact with water and the other end with gas. The miniature foaming bath that covers the surface in a glass of champagne fortunately does not contain any soap. Other molecules, such as proteins or polysaccharides, which are naturally present in the drink, play the same stabilizing role as soap molecules that slow the thinning of films.

A Geometric Stack

Let's take a closer look at these liquid walls. Unlike isolated bubbles, which are spherical, a bubble inside foam has flattened facets. To convince ourselves of this fact, just observe two comparably sized bubbles that are stuck together: their contact surface is almost flat. Why? The curvature of a bubble results from a pressure difference between its inside (overpressure) and its outside; for two bubbles of the same diameter in contact with each other, their common wall is subjected to a comparable pressure on both sides and therefore remains almost flat.

A dry foam evokes a stack of polyhedrons of different sizes, delimited by the edges where excess liquid circulates. These channels can be identified by dripping a colored drop on thick, dry foam: the drop of dye descends along the edges of the polyhedral bubbles, which are also called *Plateau's borders* (fig. 2). The capital letter recalls the name of the Belgian physicist Joseph Plateau (1801–1883), who was blinded facing the sun to study retinal persistence. His blindness did not stop him: with the help of

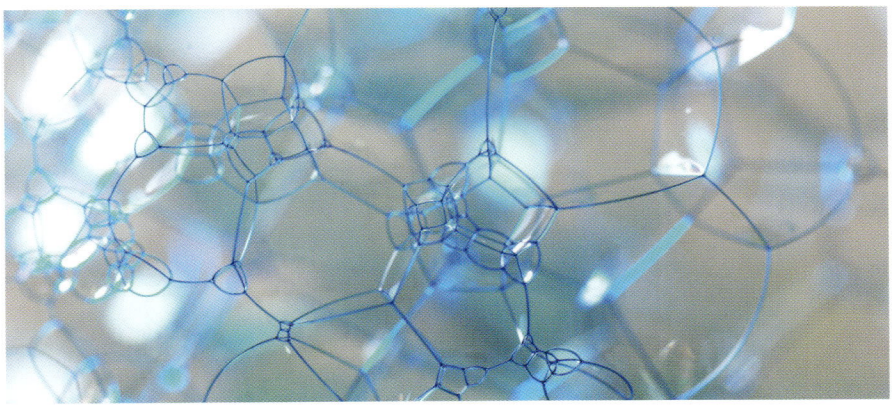

2 | The boundaries among the bubbles constitute a network of small liquid beams. Their entanglement can be visualized using foam made with colored, soapy water. The four directions of these small beams together form regular angles in space.

his son serving as his eyes, he established the laws of geometry governing the architecture of foams.

Despite their apparent disorder, there is indeed a regularity in the organization of foams. Let us focus, for example, on a single node where four polyhedral bubbles meet: notice that the edges are aligned with the four directions of a regular tetrahedron (the directions that start from the vertices and reach the center of the tetrahedron). Each edge connects three walls that meet at an angle of 120 degrees. This geometric regularity comes from the surface tension, which tends to reduce the area of the interfaces involved in a given volume of bubbles.

This particular organization of the edges is also responsible for the elasticity of dry foams. Apply very light force to shaving cream (a foam), and you will notice that the shaving cream returns to its original shape as soon as the stress is released. If the imposed deformation is great enough, however, the edges and liquid films rearrange irreversibly to form a new network of polyhedrons: some facets disappear while new ones emerge. This *plasticity* partly explains the cushioning properties of foams, their texture to the touch, and the creaminess of a chocolate mousse.

Optimal Architecture

To be complete about the architecture of foams, let's specify what building blocks they are made of; in other words, what type of polyhedrons constitutes them? Lord Kelvin (1824–1907), a pioneer in the description of the structure of foam, proposed a certain stack of bubbles that minimized the total area of all soapy walls, given bubbles of identical shape and size. Recently, Irish physicists Denis Weaire and Robert Phelan reached an even more economical, surface-area solution. They proposed piling up two kinds of polyhedral bubbles of the same volume but of different shapes. To date we don't know of a better solution! This model proved successful when it inspired the architects of the National Aquatics Center, which opened at the 2008 Beijing Olympics (fig. 3).

Let us conclude by noting that, seen up close, the facets of the polyhedrons are actually slightly curved. This curvature accounts for the low difference in pressure between two bubbles because of their size difference. Far from being innocuous, this will ultimately be fatal for any small foam bubble. Just as air can diffuse through a liquid wall, a small convex bubble, in overpressure with respect to its neighbors, gradually empties its gas; little by little, the average size of the bubbles will increase while their number will decrease. Ultimately, the foam bubbles will boil down to a single bubble that will eventually explode majestically.

3 | *The facade of the aquatic complex made for the 2008 Beijing Olympic Games, dubbed the Water Cube, is appropriately intriguing. It is paved with inflated cushions that simulate an optimal foam.*

EXPERIMENT

The three-dimensional geometry of foam is not easy to visualize. Let's keep it simple with a "flat" foam. To do this, use a straw to gently blow into a large, flat dish filled almost to the brim with water and a drop of dishwashing liquid. You'll be creating a raft of bubbles. Cover the container with a clear lid such as a Pyrex cover, so that the tops of the bubbles touch the lid (you can adjust the water volume if necessary). You'll observe paving made of mostly hexagonal polygons.

The point of intersection between polygons involves trios of edges at angles of 120 degrees. This regularity accounts for the balance of surface tension forces: the latter tend to reduce the surface of films (corresponding here to the length of the lines) by pulling in the plane's three different directions with the same force.

4

Two-dimensional bubble bed produced between two close parallel glass plates. Although the observed polygons differ in size and number of facets, their sides always meet at three angles of 120 degrees. This regularity is explained by the uniform tension that stretches each of these lines (lines that in the thickness of the cell are actually facets).

NECKLACES AND CATENARIES

What's the connection between a pearl necklace and a suspension bridge? It is their shapes, which reveal universal mathematical properties.

A row of pearls emphasizes the grace of a neck—but also the universality of mathematics! Indeed, all necklaces conform to the necks of women with the same mathematical form; of course, this shape changes if gems are attached irregularly, even though it does retain its general appearance. This universal geometrical form—the *catenary*—would be merely anecdotal, were it not also shared by railway catenaries, chairlift cables, and suspension bridges.

> **1** A necklace uniformly strung with pearls adopts a regular shape mathematicians call a catenary.

A contemporary of Isaac Newton's, the British physicist Robert Hooke (1635–1703), was one of the first to try to mathematically describe the shape of a necklace or, similarly, of a metal chain suspended between two points. Hooke was a genius inventor. The painting of him in his Oxford workspace, holding his famous chain in his hands, depicts him surrounded by several of his discoveries, notably a rudimentary microscope in the background (fig. 2). Did you know that this instrument allowed Hooke to be the first to identify plant cells? It is, incidentally, to him that we owe the word *cell*, by analogy with monastic cells.

Hooke, who was interested in everything, also played a role very early in developing the mechanics of solids, and the fundamental equation of elasticity bears his name. The memorial devoted to him at St. Paul's Cathedral in London recognizes him as "one of the most ingenious men who ever lived." Nowadays, unfortunately, there are no longer researchers with such vast knowledge, given the diversity of fields embraced by science!

A Question of Balance

Mathematically, the shape that a necklace adopts is called a *hyperbolic cosine*. This discovery played such an important role in the history of mathematics that we ended up naming a *catenary* any curve that outlines a hanging chain. Where does its particular curvature come from?

To understand this shape, let's mentally isolate one of its metal links. It is subject to its weight, but it is also pulled on both sides by its neighboring links (this tension in the chain stretches its elements imperceptibly). The two tensions acting at the ends of the link, however, are not exactly aligned, because of the curvature of our chain. The difference between these two antagonistic forces balances the weight of the link pulled downward. Thus, it is precisely the balance between tension and weight achieved at each point that determines the shape of the curve. Note that the tension in the chain is not uniform, and is higher near where it attaches.

Recent, imaginary portrait of Robert Hooke illustrating his many subjects of interest, including the famous chain.

Forces such as tension all along a catenary are general in mechanics. They are called *internal stresses*. Think of two muscular teams competing in a tug of war (the Basque game of *soka-tira*). The forces provided by the tense muscles of the athletes are very perceptible. But, as long as both teams pull with the same force (an *external* force, this time), the rope remains steady and the tensions inside the rope are balanced; only an observer with the right equipment could measure the slight variation in length of the cord element (its deformation, or *strain*) and thus indirectly estimate the intensity of the internal tension, and therefore the vigor of the two teams (fig. 3).

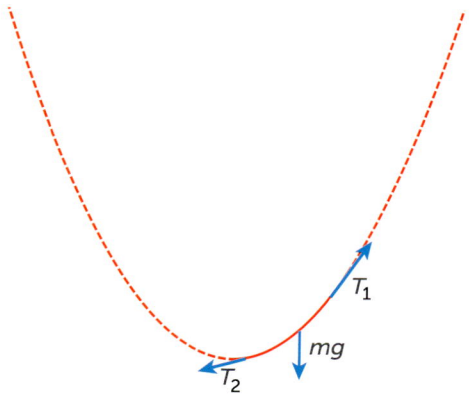

| 3 | Balance of forces on a small section of a chain (in red): the two antago- nistic tensions T_1 and T_2, tangential to the chain at the ends of the sec- tion, compensate for the weight of the latter. |

Suspended Bridges

The chain shape naturally brings to mind slender, soaring bridges. Their ancestors were the rope bridges dear to adventure movies and to Indiana Jones's adventure parks! It is not the same today because suspension bridges are designed differently: the elements of a flat deck—the part that we use—are attached to two main cables, themselves attached to the bridge's piers. Secondary suspending lines—usually vertical—connect these cables to the deck to ensure its horizontality.

The Golden Gate Bridge that extends from the city of San Francisco to the North Pacific coast is one of the most spectacular achievements among suspension bridges. Its two suspension cables evoke a necklace with small charms attached at regular intervals. But why resort to such tall piers? There's an answer available for visitors at the entrance to the bridge: to reduce the pulling force that must be exerted to stretch the ends of cables. The taller the pier, the less tension needed to support the deck.

Does the shape of the main cables correspond exactly to the catenary discussed earlier? Engineer Marc Seguin, who started building a series of innovative bridges on the Rhône river in 1823, is convinced of it. To accurately calculate the cabling needed and the piers to support the cables, he therefore used the mathematical equation of the catenary. At the same time, a bridge engineer at the École Nationale des Ponts et Chaussées, Henri Navier (1785–1836), showed that in fact the carrier cable bears the shape of a parabola if the weight of the chain is low compared to the weight of the bridge's deck—a weight transmitted at regular distances by the vertical suspension cables.

Does this constitute a quarrel among experts? The two curves are so much alike that they appear to be the same shape; a parabola is just a little more pointed than its cousin, the catenary. This is a slight but sensitive difference for builders, who seek an exact geometry of suspension cables.

Whether rope bridges or suspension bridges, the fact remains that the mechanical stability of these structures relies on the tension of their cables that must be anchored very firmly on the river banks—which, because of this requirement, are often rocky. The exception is cable-stayed bridges [INNER BALANCE] such as the recent Millau Viaduct. They make it possible to overcome this constraint because each vertical pier is acted upon symmetrically on both sides by cables.

EXPERIMENT

On the platform at the entrance of Golden Gate Bridge, an experiment for visitors shows the effect of the height of the piers on the tension of the cables. Let's create a very simplified version of it. Take an object with a handle (e.g., a cup, a jug) through which you slide a string that is about 1 meter long. Tie one end of the string to the foot of a chair laid flat on the floor. Pulling on the free end of the string, lift the object by using the opposite foot of the chair as leverage. Raising the object from the ground requires moderate effort. But, as it rises and the string nears the horizontal, the tension required increases. The weight is more easily carried if the string attaches higher up. The same is true for the cables of suspension bridges, which must rely on sufficiently tall piers to reduce tension.

4 | *The famous Golden Gate Bridge, a symbol of San Francisco, was the longest suspension bridge in the world until 1964, with a distance of 1,280 meters between the two pillars.*

Materials

ELEGANT STONE ARCHES

What, but a few ruins, remains of the mythical stadium of Olympia? The vaulted athletes' entrance. And which are the most well preserved ancient buildings? Often, stone bridges. Among these witnesses of the past, there's a common architectural principle: the stone arch.

Marco Polo describes a bridge, stone by stone. "But which is the stone that supports the bridge?" Kublai Khan asks.

"The bridge is not supported by one stone or another," Marco answers, "but by the line of the arch that they form."

—Italo Calvino, *Invisible Cities* (trans. by William Weaver)

| 1 | Bordeaux's "Stone Bridge" with its seventeen arches has been the "hyphen" that connects the north and the south of France. |

Like many architectural creations, bridges are not simply functional civil engineering structures: they decorate landscapes and may well symbolize power. Joining two areas, bridges also embody connections between people and countries. It's not surprising, then, that European bridges are featured on all of Europe's bank notes!

To build a bridge, one of the oldest and most robust solutions is to rest the deck on a set of semicircular arches—*round arches*. Arches are found in Roman work, such as the Pont du Gard, built in southern France around 50 CE, and over the centuries, throughout civil engineering, from bridges to churches. But have you ever wondered about the secret of the arches' sturdy durability?

Stacks and Abutments

To answer that question, let's first take a look at how to build an arch. Start by setting a wooden, arch-shaped scaffold between two piers, then lay the stone blocks on the arch one after the other. The word *voussoir* refers to the slightly beveled stones that ensure close contact. The structure is balanced once all the stones have been placed in a semicircle and then the wooden arch can be removed at no risk. All the stone blocks take on almost the same load: the weight of the arch itself but, especially, that of its load, which the curvature transforms into compression all along the arch.

Such compression does not harm the stones, which withstand this type of stress well; on the contrary, it guarantees stability by preventing stones from sliding. The significant compression forces are transferred laterally onto the large vertical piles that extend the vaulted arch. Without these piles, the horizontal thrust would tend to splay the arch and cause it to collapse.

Let's connect several arches. If the construction does not form a loop, as in Roman arenas, it is advisable to provide abutments (structures especially designed to withstand pressure) for the arches at the ends. In the case of a bridge, these abutments rest against the banks of the waterway.

Bridges are nothing special when it comes to systems of arches: from the start of antiquity the principle of the vault has been applied to dwellings. The roofs of houses then were often made of wood or thatch, both highly flammable materials. Stone blocks would eventually be used, but one question remained: how to support this heavy, rocky roof at the entrances?

2 | *The entrance of Olympia stadium, where the Olympic Games were first held in the seventh century BCE. The beginning of the vaulted way is still intact.*

<table>
<tr><td>3</td><td>*Perspectival view of the cloister of the Cahors Cathedral and its rich set of pointed arches. Leading back to pillars, they allow numerous openings onto the cloister, all the while supporting a high ceiling.*</td></tr>
</table>

An initial solution was to use long, horizontal stone blocks—*lintels*—above windows and doors. However, this solution soon revealed its limits in large buildings, such as temples. Indeed, the Egyptians and the Romans quickly realized that rectangular blocks poorly resisted deformations caused by stress because of the flexion created by their weight: they cracked on their lower facet. The use of lintels therefore required that the vertical uprights supporting them be close enough to reduce their cracking. That's why the columns of the famous Parthenon are no more than 5 meters apart. Despite this precaution, some lintels have split over the centuries and have had to be rebuilt in modern times. In contrast, the archway at the entrance of the ancient Olympia stadium has survived intact (fig. 2). It seems incredible that an isolated stone arch has held up to the present day, but there is a clear reason for this. Whether it supports a bridge or roofing, the arch guarantees greater strength than any vertical wall! This sturdiness comes, as we have seen, from the fact that all the stone elements that compose it are compressed.

A Church's Vaults

The counter-intuitive sturdiness of an arch contributed to the development of this construction technique after antiquity, especially in Christian churches, establishing an endless dialogue between forces and shapes. The Romanesque style would first light the interior of religious buildings by creating semicircular arches—or *barrel vaults*—resting on thick, vertical pillars. But this robust semicircular shape limits the height to the radius of the vault, which is to say, to the spacing between pillars.

In the Gothic period, this shape would gradually be replaced by pointed (or Gothic) arches that meet at the top at a *keystone* (fig. 3). With respect to the semicircle, the pointedness of the arch makes it possible to increase its height, thus permitting more light to enter, often filtered by stained glass windows.

The invention of *ribbed vaults* (with crisscrossing arches) seeks to concentrate the weight of the upper layers on pillars that thin out and lengthen, opening wide spaces between them. But these slender columns cannot balance the transverse forces alone. To remedy this, architects equipped their structures with *flying buttresses*, arches that transfer lateral pressure to abutment pillars. The flying buttresses that elegantly adorn the exterior of cathedrals like Notre-Dame de Paris are not merely aesthetic. This balance of forces that rely on the columns will also be the primary function of the lateral galleries.

The goal pursued in building cathedrals during the five centuries of Romanesque and Gothic art can also be found in Islamic art. Here, too, architects sought to create maximum space for the many faithful and for the structures to soar. It's all about reaching for the Most High!

EXPERIMENT

How does one demonstrate the strength of an arch? Thread a necklace tightly using large beads. Nail the necklace ends to a wooden board so that the beads on both ends touch the nails. The necklace adopts the catenary shape described earlier [NECKLACES AND CATENARIES]. Without making any change to the shape the necklace naturally assumed, gently tilt the board along a horizontal axis. When the board is flat, the chain shape remains the same, which is hardly surprising.

But now continue to gradually tilt the board: the inverted vault remains stable! The distribution of forces that ensured the shape of the necklace is simply reversed. What was tension now becomes compression, and the absence of transverse forces ensures the equilibrium of this model vault, which holds up without flying buttresses. If you change the shape of the arch, the structure will probably collapse, because the inverted necklace cannot withstand transverse forces that would slide the beads on top of each other. The architect Gaudí used the inverted-chain shape a great deal in his design of Barcelona's famous cathedral, La Sagrada Família.

As hangs the flexible line, so but inverted will stand the rigid arch.
—Robert Hooke

Materials

1.

2.

3.

4.

5.

SHELLS AND
MILLE-FEUILLES

The interior of some shells is covered with mother-of-pearl, a
material with exceptional mechanical properties. The secret of its
extraordinary hardness is its microscopic structure, similar to a
stack of bricks separated by a thin layer of glue.

---

*[The] conch [grows] by development of turns. This shows that everything, in
which form and matter are linked and preserved one by the other, has its time.
The form is then an integral of time which absorbs, as it were, the preceding
ages into the present.*
—Paul Valéry, *Cahiers*

A beach at low tide. Innumerable periwinkles, cockles, mussels, and other
shells flourish amid the rocks. A curiously shaped shell stands out from
the crowd: it looks like an auricle that one might have pierced with a

1	*An abalone shell forms a lobe pierced with a row of small holes. Its interior is covered with iridescent mother-of-pearl of extraordinary mechanical toughness.*

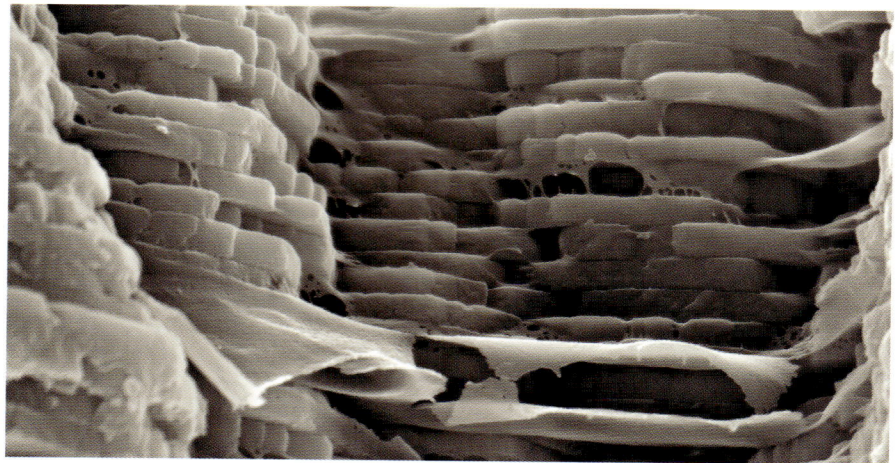

2 | *In mother-of-pearl shells, half-micron-thick aragonite platelets pile up into a wall of microscopic bricks. A thin layer of soft mortar unites them. This assembly confers surprising strength to the material.*

dozen holes drawing a graceful arch. It belongs to an abalone, that mollusk [that French gourmets call Breton caviar or "*caviar de Bretagne*"] whose shell is adorned with magnificent iridescence thanks to the mother-of-pearl covering its inside. But it has another unexpected and even more remarkable property: it is incredibly durable, much like the most robust steel!

Well-Cemented Microbricks

Mother-of-pearl is the source of the abalone shell's sturdiness. Many kinds of shells produce this shimmering material that is used in inlays, jewelry, and decorations. It owes its iridescence to interference of light reflected by superimposed layers of *aragonite*, a crystalline variety of calcium carbonate akin to chalk. More precisely, an electron microscope image reveals that aragonite is organized in a stack of thin tablets, in the manner of a brick wall (fig. 2). The thickness of these platelets is about half a micrometer, on the order of the wavelength of light. Chalk is a very brittle material, but what is more paradoxical than the extraordinary mechanical toughness of mother-of-pearl, 95 percent of which is aragonite? In fact, it is the microscopic arrangement of the material that confers toughness to the shell. The polygonal crystalline slabs are separated by layers of a soft mortar consisting of polymers, mainly proteins and chitin (a molecule of the carbohydrate family). This binder, which is no more than 5 percent of the mother-of-pearl, forms films still twenty times thinner than the platelets themselves. And yet, it is largely what makes the layered structure so strong. Let's see why.

The Mille-Feuille Strategy

The breakage of an object usually stems from a flaw on its surface, which then propagates through the whole material. For some materials such as glass [GLASS TEARDROPS], the crack progresses extremely rapidly and presents smooth facets, or fracture surfaces: that is a *brittle fracture*. In contrast, if you stretch a piece of modeling clay, it will stretch out a great deal and irreversibly before breaking: that is a *ductile fracture*.

In the case of abalone, breaking the shell means propagating a crack through the wall of microscopic bricks. If you look at it very closely (fig. 2), however, mother-of-pearl has many growth defects that could act as crack initiators. Why is the shell so resistant?

In fact, the aragonite platelet complex strongly resists crack propagation. A crack in a hard plate of aragonite will soon encounter a soft layer of cement, where some of its energy will dissipate. Consider how difficult it is to cut a mille-feuille: the force exerted on the knife through a hard layer of puff pastry is brutally redistributed by the creamy soft layer, leading to disaster (fig. 3)!

The mille-feuille disaster—layers sliding around relatively to each other—could also occur in a shell. But, in order to contain the disaster,

3

Cutting a mille-feuille cleanly is a challenge because the knife alternates between encountering puff pastry and cream. Nature uses the same type of structure, an alternating stack of hard and soft layers that gives mother-of-pearl its extreme toughness.

nature has provided an original answer: the organic binder adheres very strongly to the aragonite slabs, preventing their separation. This binder is elastic and can take a great deal of strain without breaking. Finally, the surface of the crystalline platelets is marked by nanometric roughness. These small nubs ensure a very strong anchoring of the bricks in the surrounding organic mortar. The rough surface of one layer commonly almost touches the facing layer. Their presence drastically hinders relative slippage and inhibits the separation of the aragonite layers. Voilà!

Mimicking Mother-of-Pearl

Mother-of-pearl, a marvelous natural material, combines the stiffness of aragonite bricks with the softness of a polymer mortar. But it is not the only composite material to combine the characteristics of its two components to achieve outstanding mechanical properties. That is the case of wood, tooth enamel, and bones. Mother-of-pearl is precisely biocompatible with bone, making it a valuable bone substitute.

Nonetheless, the availability and variability of these natural materials make it difficult to use them directly. This is why engineers have adopted a process of copying, or rather of being inspired by natural materials whose structures and properties are transposed. This *biomimetic* approach, though very attractive, is not so easy to implement and does not work all the time. And, although you can grow a crystal in the lab, growing a shell from its natural ingredients is a challenge.

Soft Chemistry

Biomimicry is one of the current research challenges in *soft chemistry* as one of its experts, Jacques Livage, has called it. How does one synthesize strong materials from an abundant material such as limestone at room temperature? These syntheses are usually achieved at high temperature (1200 degrees Celsius to make glass) [STATES OF GLASS]. Meeting this challenge can lead to considerable energy savings.

Sol-gel (i.e., solution-gelation) processes make it possible to go from a liquid state, made of a solution of particles in suspension, to a crystalline network when these particles react with each other to form a solid network whose solvent will later be eliminated. Such processes are used in nature [BUILDERS], and current research is trying to mimic the production of such vitreous materials, which occurs at near-ambient temperature.

EXPERIMENT

Surprisingly, the toughness of mother-of-pearl and, more generally, of a composite is related to the fragmented nature of the material. Let's study this effect with the packaging film of a box of cookies.

Cut out two rectangles A and B of this film, about 5 cm × 15 cm (1).

A will be the reference sample (2).

Cut parallel fringes lengthwise, in rectangle B (3).

Cover these rectangles completely with a wide strip of packing tape (4) to obtain two samples that are identical in appearance.

The experiment consists in trying to tear the rectangles widthwise, starting from a notch made with a pair of scissors (5). Tearing rectangle A should not be a problem; on the other hand, cutting B will be tougher. You will be able to tear the first fringe without any problem, but watch how the tear barely spreads across this model of composite structure.

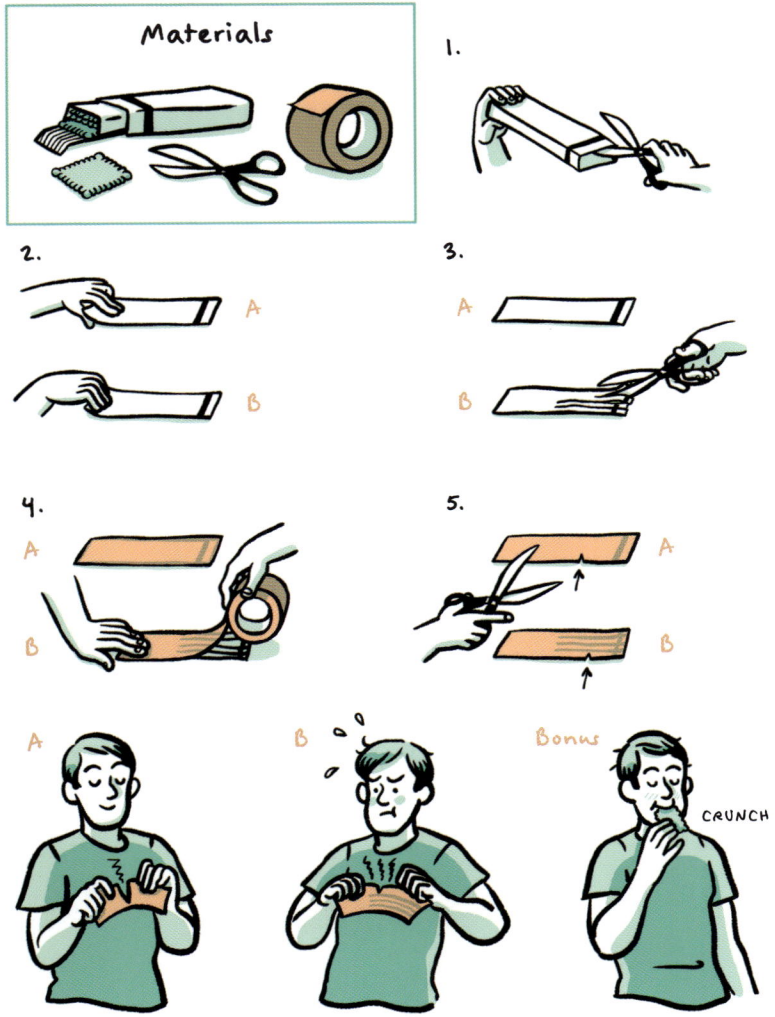

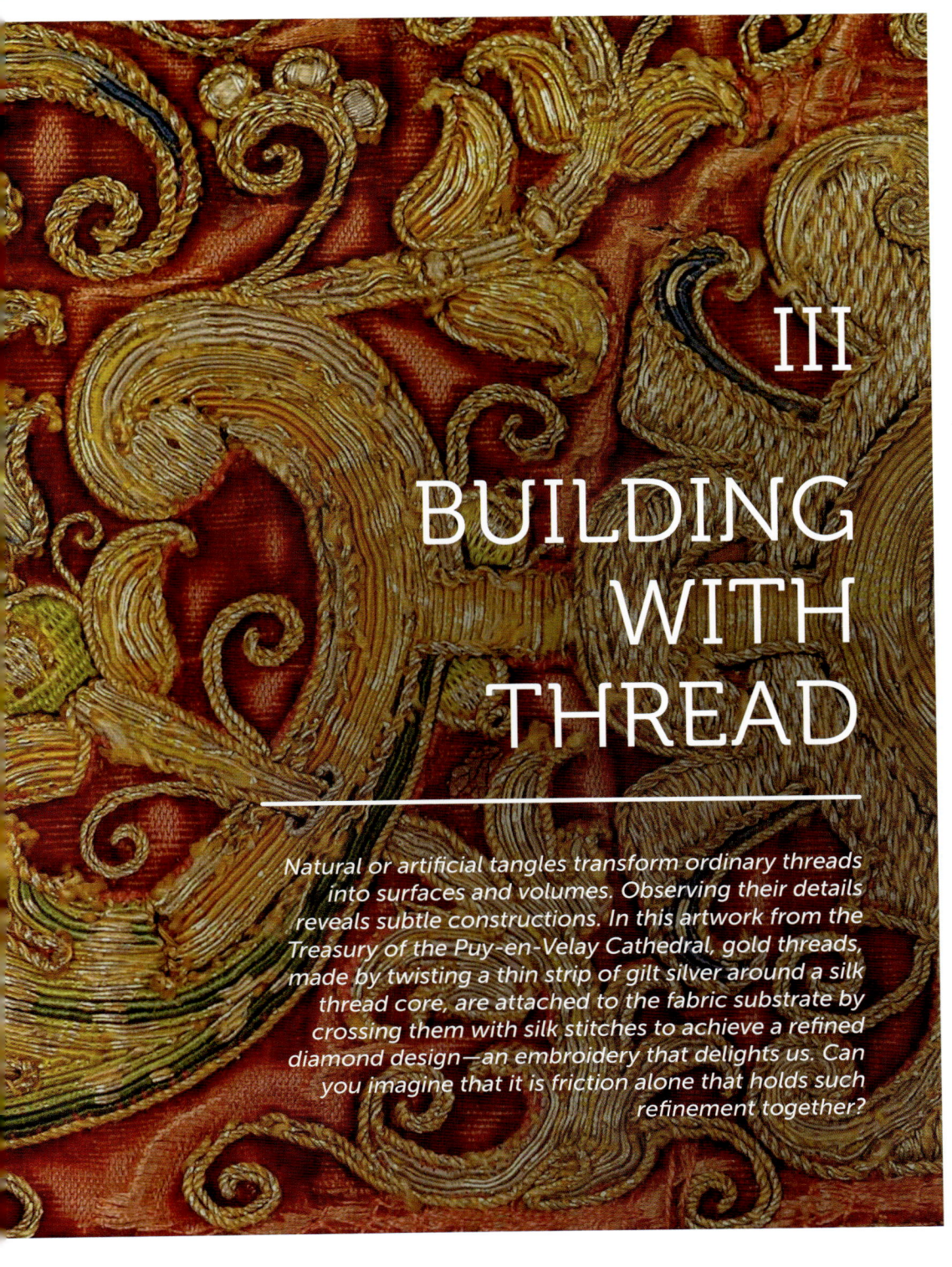

III

BUILDING WITH THREAD

Natural or artificial tangles transform ordinary threads into surfaces and volumes. Observing their details reveals subtle constructions. In this artwork from the Treasury of the Puy-en-Velay Cathedral, gold threads, made by twisting a thin strip of gilt silver around a silk thread core, are attached to the fabric substrate by crossing them with silk stitches to achieve a refined diamond design—an embroidery that delights us. Can you imagine that it is friction alone that holds such refinement together?

AN EIGHT-LEGGED BUILDER

What's more common than a spiderweb? And yet it's capable of an extraordinary feat: it can stop an insect traveling at 40 kilometers per hour without the insect bouncing off or breaking the web. Where do the spiderweb's unique, mechanical properties, which could revolutionize the world of textiles, come from?

A few drops of dew on a spider's web, and there you have a diamond necklace.
—Jules Renard

On an early morning walk, who has not admired the silk patterns patiently woven overnight by the common garden spider and their beaded wreaths of dew? The archetypal spiderweb is a radiating, star-shaped, cabled structure that anchors itself strongly to neighboring branches. Have you noticed how these cables are connected thanks to a secondary, transverse thread forming a spiral? So many all-important details because the spider's meal will depend on the prey it has captured in this formidable trap. Silk seems fragile, but in fact, its resistance compares to steel's. Understanding its production would be invaluable for the synthetic textile industry.

| 1 | *Decorated with dew drops, this spiderweb combines stiff radial cables that support it and extensible circular threads that will trap a rash insect.* |

As Strong as Kevlar

Radial and spiral cables do not have the same mechanical advantages. The former offer a strength that remains out of reach of most synthetic polymers. If you have the opportunity to travel to the tropics, you may encounter the tropical kin of our garden spiders, the golden silk-orb weavers from the *Nephila* genus. Nephila spiders are relatively harmless despite a span of several centimeters, and the silk they produce breaks every record of strength: the cabling is six times lighter than steel, and it has the same breakpoint threshold: a silk cable with a 1-square-centimeter cross section could thus support a truck weighing 10 tons!

This remarkable strength occurs as a result of silk's molecular structure. It is composed of proteins that combine elements of very rigid, crystalline structures, and amorphous parts that give the molecular structure some flexibility. Only well-known synthetic Kevlar, discovered by American chemist Stephanie Kwolek in the 1960s, rivals the silk of these tropical champions.

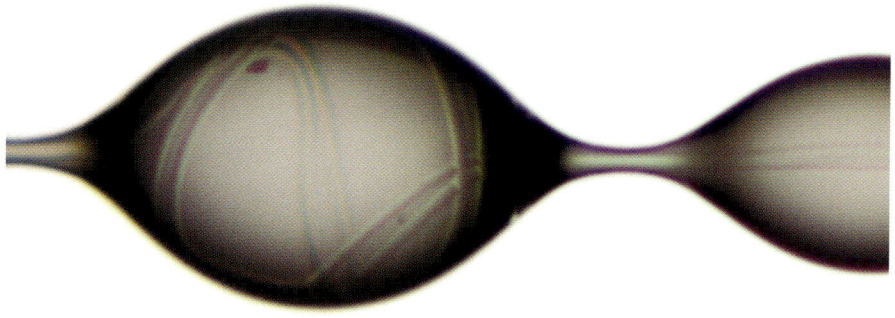

2 *Excess fiber is wound inside the globules that are one-tenth of a millimeter in diameter and distributed along the transverse threads. When such a thread is stretched, its fiber unwinds without increasing tension.*

Shock Absorber

The spiraling silk threads of the spiderweb present unparalleled elasticity: if you pull on the web, the fibers can reach three times their original length before breaking. It then takes a few seconds for them to regain their original length. But why don't they behave like radial cables? The role of a spiderweb is obviously to capture prey—which requires holding on to a bee, for example, which may be flying at 40 kilometers per hour at impact. It can't be only strong: a *trampoline* web would not secure the spider's dinner. Any contact must be efficiently cushioned to prevent the prey from bouncing out of the predator's reach.

Recent experiments conducted jointly by physicists and an arachnologist have revealed an amazing mechanism that, for many spider species, converts impact into a soft collision. Look closely at a spider's web: with a little bit of focus and luck, you will detect small, viscous globules distributed evenly along the capturing threads the spider has woven—not to be confused with that morning's condensation of dew drops.

What purpose might these globules serve? Researchers have shown that they store a large excess of spiraling thread that unspools at impact (fig. 2). The capillary forces, which control the cohesion of wet sand [THE SECRET OF SANDCASTLES] and the bonding of locks of wet hair [WET HAIR], assure the initial tension. The globules act like shock absorbers: thanks to them, a prey will not bounce out, and instead becomes entangled and quickly trapped.

A Staggering Jelly

But how do spiders produce silk? Their abdomen contains glands shaped like elongated light bulbs where they make a protein jelly. This amazing fluid, whose viscosity is several million times greater than that of water, is carried along a channel that gets progressively thinner as it forms a nozzle called a spinneret. Its diameter thus goes from being a fraction of a millimeter long to a few tens of micrometers, a scale much smaller than the diameter of a hair. What mechanism enables such a viscous fluid to flow down such a narrow channel? There's a mystery! Even more surprising is what happens to the jelly along the way: a physicochemical miracle progressively transforms an aqueous protein gel into a rigid and water-insoluble fiber. How this miracle happens remains a largely open question.

Four hundred million years of evolution have endowed the order of *Araneae* with an incredibly complex internal machinery. Using their seven pairs of spinnerets, some species are able to produce silks of different compositions, depending on the function they need: radial cables that ensure the strength of the fabric; spiral-wound strings that dampen the impact of prey and capture it; or special thread for swaddling future victims. Each fiber has its particular characteristics.

Tomorrow's Textiles?

We've known of the remarkable properties of spider silks for a long time. The Greeks used them to make bandages and even as suture thread. The Papuans of New Guinea also made landing nets by reusing spiderwebs on a frame made of branches. Beyond these ancestral techniques, astronomers made use of the refinement and strength of spider silks for the reticles of their eyepieces. Who knows? The discovery of some stars may have occurred thanks to a mere thread of silk!

Why not use spider silk to make fabrics, like the silk of the mulberry worm? In the middle of the nineteenth century, the explorer Alcide Dessalines d'Orbigny (1802–1857), visiting South America, recounted the use the natives made of spider silk; they are said to have even made him some comfortable silk pants. A bit later, Father Paul Camboué (1849–1929), a Jesuit missionary in Madagascar, sought to collect the golden silk of the good-sized local species *Nephila Madagascariensis*, the golden-orb silk weaver capable of producing 4 kilometers per month of a silk that can't help but remind us of the golden thread.

This work led to the making of a bedspread introduced at the Universal Exhibition in Paris in 1900. The technique fell into oblivion until artists

Simon Peers and Nicholas Godley accepted the challenge of creating a golden spider-silk cape. This huge project took years of hard work and over a million spiders to get 1.5 kilograms of silk. In sum, over seventy people were involved in collecting, spinning, and weaving operations. The result is a magnificent work of art on exhibit in London (fig. 3). And yet, any industrial adaptation of the process is still utopian. One possible path would be to reconstitute fibers from the proteins that spiders produce. But we are still far from matching the tour de force achieved in the spinnerets of our garden guests!

3 | *This Nephilia spider-silk cape is on display at the Victoria and Albert Museum in London.*

EXPERIMENT

The droplets that contain little balls of excess silk as described previously are often superimposed on regularly spaced dew drops. Here is how to reproduce such a string.

Unfold a paperclip into an L-shape (1). Dip the base of the L in a bath of cooking oil and remove it while keeping it horizontal (2) and (3): a sheath of liquid covers the metal (4).

A few seconds later, you will see that this sheath gradually destabilizes and becomes a string of drops (5). Surface tension forces are known to always seek the minimal surface area of a liquid [SURFACE TENSION]. From the perspective of surface tension, the spherical shape of the drops is much more favorable than the initial cylindrical interface they start with: it presents less surface for the same volume. The radius of the wire controls the regular spacing between any two drops: the finer the wire, the smaller the distance between the drops.

Materials

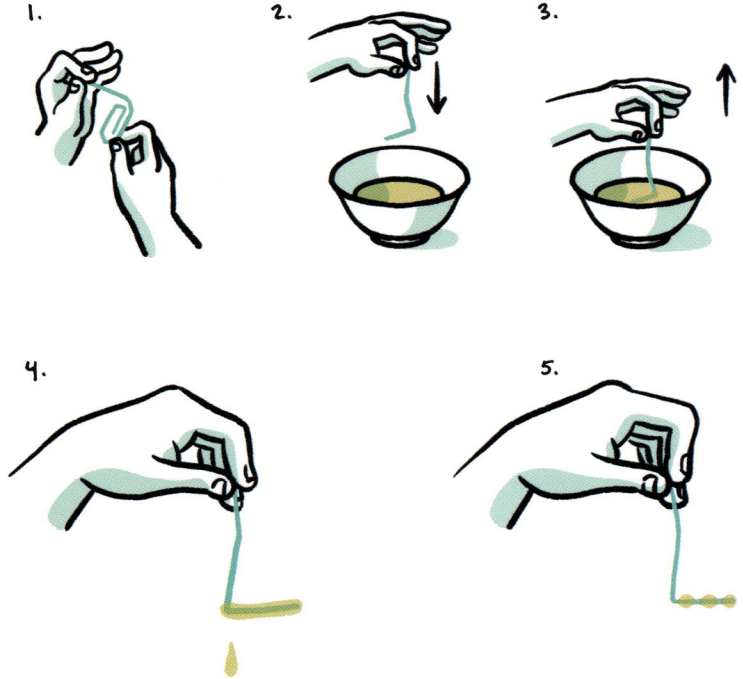

WET HAIR

Have you ever taken the time to see the surprising figures that result from the arrangement of hairs when your hair is wet? They are produced by the mechanical stiffness of fibers arm-wrestling with capillary forces. Such an unexpected, mundane process is essential to how microsensors operate in our mobile phones.

A priori, the early twentieth-century actor Rudolph Valentino and the punks of the 1980s subculture have nothing in common—nothing, that is, but their obsession with hair. The former remains famous for his slick, dazzlingly luscious hair; the latter for their Mohawk hairstyle straight out of a western movie. Gels or just water have long been used to battle

| 1 | *The hair stands up and organizes into locks, thanks to capillary sticking, using water or a styling gel.* |

the common disorderliness of hair and enhance beauty. But have you ever paid attention to how wet hairs gather together? They never line up nicely next to each other, as one might expect.

Experiment with someone around you—ideally someone with a crew cut or spiky hair—coming out of the shower. Watch their hair rise starting from the tips. You will see that the hairs do not appear individually, but are organized in locks, themselves divided into sublocks, and so on to the roots. What brings about this strange and spontaneous branch structure?

Styling Hair in the Lab

In order to study the agglutination of hair under controlled conditions, researchers have devised an experiment: dip a sort of artificial brush made up of thin, regularly spaced strips into an oil bath (fig. 2). Once removed, the brush forms a most elegant tree figure. At the very top, the strips begin to come together, via a liquid bridge, in pairs. This process of association is repeated to produce thicker and thicker strips.

Surprisingly, the experiment shows that the size distribution of the locks is independent of hair length. In other words, this experiment suggests there's a universal way that wet hair gathers.

The cascading process that generates these slender structures is more general: it is what physicists call the phenomenon of *coalescence*. Basically, small objects combine with each other to form larger objects, which

2   *In this laboratory experiment, a set of parallel plastic strips held at equal distances by blue, 1-millimeter shims was immersed in a bath and then slowly lifted out. Liquid bridges (which appear dark) are created between the strips intriguingly organized like trees.*

themselves associate in turn to form larger objects, and so on. The same phenomenon is at work in a cloud, where the association of tiny droplets will eventually lead to a downpour.

The Right Balance

What are the physical ingredients driving this aggregation? No need to tear your hair out! Here is the answer: it is all a showdown between the attraction of the strips and their mechanical stiffness. As for sandcastles [THE SECRET OF SANDCASTLES], water plays an adhesive role. Just as a sandcastle drowned by the tide falls apart, so too will too much water damage your hair's cohesion. Thus, a tousled paint brush will remain so if soaked completely under water, because there will no longer be any interface between water and air. It will only take on a nice, sharp shape if it is damp, which is to say neither dry nor totally wet.

Getting our hair to stick together, however, requires that hairs bend, which goes against their mechanical rigidity. Thick, stiff hair is less apt to bend and will produce fewer strands than thin, flexible hair. If our hair is too long, say beyond a few centimeters, a third ingredient becomes the spoilsport: the weight of the fibers will outweigh the strength of the structure. By greatly increasing adhesion and mechanical rigidity, a styling gel nevertheless makes it possible to maintain uncanny peaks about twenty centimeters long!

Threat in Mobile Phones

But why do the researchers split hairs and quibble about wet hair? In fact, hair's self-assembly mechanism also occurs in the tiny mechanical systems of many everyday apparatuses, such as the device that informs mobile phones of their orientation with respect to the vertical direction. This sensor has flexible structures as thin as a human hair in the form of tiny combs: the teeny teeth of these tiny combs bend under gravity or acceleration, and the electrical measurement of their deflection indicates the phone's tilt (fig. 3). An innocent drop of water, condensed from ambient humidity, is enough to impede the sensor's proper functioning, because that drop would stick the strips to the substrate.

To guard against such effects, *micro-engineers* must learn to master the capillary forces at play in their inventions. But for whoever knows how to tame these forces, it is also possible to make use of them wisely. Have you ever been impressed by the multitude of tiny components soldered to a circuit board? In an industrial production line, an automatic soldering iron deposits drops of solder at points of contact. A fraction of a second later, a robotic arm roughly places a sort of centipede—an electronic component—on these drops. The liquid solder attracts the legs to the contact points and adjusts the component's position. The latter fixes quickly as the solder hardens. Current research seeks to go further in the use of capillary forces to develop small-scale, three-dimensional struc-tures—a sort of *capillary origami*.

3 | An accelerometer's double-comb structure informs a mobile phone of gravity's direction and provides information about its orientation. One of the combs is fixed, and the other shifts slightly under the effect of its weight or, more broadly speaking, if it is shaken. The width of the image is comparable to the diameter of a human hair. Therefore a drop of water is the enemy of such a device: sticking the combs together prevents the device from operating properly.

4   *A capillary origami: a soap bubble was placed on a very thin flat sheet cut in a pattern of a flower. The structure then closed up under the effect of capillary forces between the bubble and the sheet (see p. 102). The same phenomenon occurs on a small scale when a drop of water is deposited on a flexible-enough sheet.*

EXPERIMENT

A very elegant but delicate experiment consists in folding a flat sheet into a three-dimensional object (capillary origami) using a bubble.

Cut a 5-centimeter-sided, equilateral triangle out of a very thin emergency blanket (1). Put it on paper towels.

Brush the upper side of the triangle with soapy water, avoiding overflowing (2).

Using a straw, try to create a bubble at the center of the triangle (3). The bubble often bursts, but with patience you will get a beautiful pyramid. If you are even more ambitious, you can cut out a flower pattern and shape the bud shown on the previous page.

1.

2.

3.

4.

pff

pff

103

BIRDS AS ARCHITECTS

Nests come in a wide variety of shapes depending on the species, with smaller birds often building the most elaborate shelters. What is the secret of creating these small, architectural wonders? The answer: friction among twigs, as with pick-up sticks!

A male weaver sports a beautiful bright-yellow chest; he also bears a black mask that makes him look like a winged Zorro. Contrary to what an unobservant tourist might conclude, this inhabitant of the African savanna is not known for his flattering appearance, but for his particular, seductive behavior.

> **1** *This weaver makes his nest by interweaving twigs. The result is strangely reminiscent of weaving or basketwork. What holds this construction together? The friction among the fibers, that's it!*

When mating period arrives, he builds a nest in a tree or in reeds, then stands at its entrance and begins to display: he flaps his wings and emits calls to attract a female. She will then take over the nest where the birds will soon mate. The polygamous, masked weaver will build several nests to house his harem during mating season. It is clear that, in terms of sexual selection, his major assets are neither strength nor beauty, but his talents as a builder!

For whoever observes them closely, nests are wondrous objects, and birds are remarkable architects. In his *Natural History*, Pliny the Elder already invited us, in a biomimetic approach, to be inspired by these winged builders. The masked weaver's nest is a cocoon that hangs from branches; it is so well woven that it holds on with just a little bit of grass.

Building a Thrush's Nest Step by Step

To better understand how birds perform their architectural wonders let's follow the work of the familiar thrush step by step. Although it uses its precious nest for only a short time, it takes great care building it. Judge for yourself: it starts by positioning twigs a few centimeters long at a more or less horizontal branching of a tree, in order to set its nest. It crisscrosses twigs and small branches one after the other around an empty hole. Once this scaffolding is completed, blades of grass and leaves attached around its upper area guarantee both the coherence of the whole and effective camouflage. Finally, the bird carpets the bottom of its nest with moss or mud to stiffen it and plug any remaining holes.

Thrushes are opportunistic. Indeed, they use materials they find around them: that means the thread of a spiderweb or even a cocoon will guarantee the nest is well ensconced in the tree that hosts it (fig. 2). Such threads even become basic elements in more evolved nests, where they are the support and attachment structure. By pivoting inside a nearly finished nest, a thrush ensures its hemispherical and gentle shape: it is ready for laying eggs and nesting. This is a very general type of nest and

2 | *Thrushes establish their nests at branchings. They lay small, intersecting branches and flexible stems that fold to shape a cup. They finish it off by adding moss and other soft debris.*

is found among aptly named bee-hummingbirds: by reproducing the same operations at their scale of a centimeter, these birds build nests that are hardly bigger than walnut shells!

A Masked Basket Weaver

Whereas the thrush is ingenious in appropriating materials around it, the masked weaver is ready to do anything to create real basketry. First, it knots blades of grass around two solid twigs to make a kind of hoop. Have you ever tried making a knot with one hand? Well, the weaver does it with just its beak. Using natural threads, it then weaves a large pocket with a roof, a floor, and walls whose small openings bar predator access. It usually takes a day to build the hanging basket in which chicks can hatch and grow.

Depending on the species, there is a great variety of construction techniques. Swallows are expert masons: they build their nests by aggregating mudballs. Malay swiftlets build their nests with their saliva: in parts of Asia, these nests, dissolved in a soup, are a delicacy. As for the award for the most decorated nest, that redounds to the bowerbird from Australia and New Guinea, which is a distant cousin of the facetious jay and thieving magpie. In order to impress the Missus, the Mister develops elegant twig arbors, adorns their entrances with flowers, berries, and multiple colored objects, whose sources are unfortunately often industrial.

Class Dunces

But not all birds are so active. If there are master architects, some lazy ones are reluctant to lift a beak, as it were! Cuckoos are squatters, and their eggs sneakily benefit from hiding among the hatchlings of another species. As for ostriches, they are satisfied with a simple hole at the soil's surface, in which they bury their eggs. And for woodpeckers, some cavity in a tree trunk is amply sufficient.

For lack of available materials or because they don't need a complete nest, other species build dwellings that are nests only in name. Penguins thus shape circular buildings by means of pebbles—real trophies that they don't hesitate to pilfer from their peers. As for eagles, they are satisfied with a nest, or *aery*, consisting simply of large, entangled branches: having no direct predators, large raptors have indeed no need to hide!

Construction Secrets

Beyond the elegance of nests, what are the physical mechanisms involved in building them? How can a complex assembly of twigs and fibers stand? Certainly birds sometimes use binding elements such as mud, saliva, spiderwebs. And yet the strength of the assembly essentially stems from friction among its elements. Remember the game of pick-up sticks, and how careful you have to be to remove a stick without displacing other sticks, especially when they are all piled on top of each other. Stalks press against each other at a point of contact, under the weight of the stalks above them. Friction stabilizes the whole by reducing the stems' sliding around. The use of flexible or branched stems amplifies this effect by multiplying contact points and geometric blockages, as in renowned architect Wang Shu's "Garden of Lingering Clouds" (fig. 3).

The most effective way, however, is to braid the stalks. In this case, the pressure induced by the intertwining considerably increases the intensity of the frictional forces at contact level [AN IMPRESSIVE BRIDGE OF GRASSES].

Aegagropilas

There is another way to understand why the phenomenon of friction alone is enough to consolidate small construction: by taking a walk on the beach. You may have spotted these strange, oblong balls; they look like kiwis. These *aegagropilas* are made of small short fibers that can be removed one after the other from the surface (fig. 4). They are fragments of aquatic plants whose meadows are vital to many species that spawn along the coast. Gathered randomly at the will of waves, the individual fibers intertwine, forming a compact ball whose size grows gradually until the ball runs aground on the beach.

3 | These pieces of wood piled on top of each other at the "Garden of Lingering Clouds" at Chaumont sur Loire are the work of landscape architect Wang Shu. They evoke an inverted bird's nest. It remains standing thanks to the friction among the beams.

How do these fibers become entangled? It may be that the stalks line up along the lines of a current and roll up in the water's whirl. There are French research teams interested in understanding this mechanism using model experiments. This seemingly futile problem is indeed crucial in many industrial applications, such as rainwater management or the clogging of a washing machine!

4   *On some beaches, you can find tightly woven balls—they look like nests—that only hold together thanks to the entanglement of and friction among fibers. The aggregation mechanism that occurs when they roll around in the waves is still poorly understood.*

EXPERIMENT

How do you lift a hefty directory using another directory without setting one on top of the other? Here is a simple and surprising experiment that illustrates the extraordinary power of frictional forces. (Use large-format paperbacks for this experiment if you can't put your hands on old-fashioned telephone books.) Create a paper mille-feuille by inserting some of the sheets of each book into the other, ten times or more (1) and (2). Now try to separate them. Hercules wouldn't be able to do it!

Where does this colossal strength come from? By interleafing the sheets, we add frictional forces. Even more, this entanglement amplifies the friction between the batches of sheets by considerably increasing the load that keeps them touching each other. It is a similar mechanism that makes the weaver's basket-like nest so resilient.

AN IMPRESSIVE BRIDGE OF GRASSES

Would you cross a steep gorge on a bridge made of spun and woven grass? An incursion in the physics of the capstan will be enough to reassure you regarding friction's superpowers. It is indeed friction that turns a grouping of short fibers into a long and particularly robust rope.

Every June in the Andes, two communities separated by the mighty Apurímac River meet for three days to rebuild a 130-meter-long rope bridge, according to the ritual followed by their Inca ancestors. This ecological, participatory bridge is built from grasses about 50 centimeters long. The villagers scythe the grasses and then beat them to extract their fibers, as with flax. A cascading construction using fine, short fibers produces a long and strong rope.

| 1 | The Q'eswachaka Rope Bridge in Peru. At the bottom right are the remains of the old bridge. |

Once the fibers are created, they are assembled, spun, enmeshed, and interwoven with each other. The frail yarns thus made are then gathered into cords, which are in turn gathered some thirty at a time and spun into long cables. Finally, the cables are joined three at a time, braided, and stretched by the strongest men of the two villages (fig. 2). These braids, whose diameter measures about 15 centimeters, serve as the base of the floor and the rest of the bridge's structure. They are pulled across the canyon by means of the old bridge, which is eventually detached and recycled, so to speak, swept away by the waters of the Apurímac.

A Mechanical Challenge

If you can't travel all the way to Peru to uncover the secrets of making ropes, maybe the Corderie Royale de Rochefort, on France's Atlantic coast feels more accessible. It details the minute operations required to make ropes for ships in the French Navy. This grandiose building, some 375 meters long, was built in the seventeenth century. Used for nearly two hundred years, it is now a museum. There one learns in particular that, until the advent of synthetic fibers, huge ropes were made of early strands of hemp, about 10 centimeters long.

Thus, the ties of suspension bridges, just as their nautical equivalents, raise the same question from a mechanical perspective: what is the magic of spinning that transforms short and fragile plant fibers into long, robust cables? We know that manufacturing yarns from vegetable or animal fibers for making cotton and wool clothes is a very old practice. From spindles invented six thousand years ago to spinning wheels dating from the Middle Ages, spinning has long been a part of human culture. It was even a symbol of passive resistance against the British occupation of a young country, India, by Mahatma Gandhi, who wove his own cotton for his clothes.

2 | *Final weaving of the long bridge cables. Each strand comes from the assembly of cords, themselves created by spinning humble, dry grasses.*

Da Vinci and the Capstan

Spinning combines fibers previously arranged by combing fine and long fibers, or by *carding* irregular fibers (carding is a kind of combing, done using a flat instrument spiked with nails). To *spin* is to amalgamate fibers and pull them into a continuous strand using a movement that combines traction and twisting. How do these contortions make the obtained yarn resilient? To answer this question, let's take a look at some naval gear: a *capstan*. A capstan is a cylinder used to maintain the tension of the rope wrapped around it. The more rope is wound around the capstan, the less force needs to be exerted to hold the free end of the rope. In practice, sailing enthusiasts can see that wrapping the rope two or three times around the winch is enough to control the jib sheet with minimal force. But release this force completely, and the rope starts slipping inexorably.

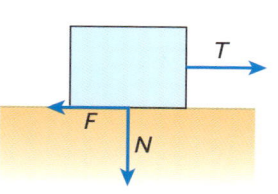

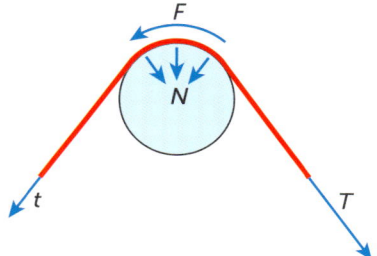

3 A block placed on a plate (left) slips only if the tensile force T is greater than the product of the perpendicular force N applied to the block (its weight for example) by a certain coefficient. The latter accounts for the static friction between the surfaces that touch.

In a model capstan (right), the strand of rope stretched on a cylinder also undergoes sliding resistance F, which is the greater when tension T exerted by the boat is stronger: indeed, the rope's tension is accompanied by a normal force that presses it against the cylinder. A small, tensile force t is thus enough to hold back the strand subjected to high tension T.

To explain this effect, let's return to an experiment dear to Leonardo da Vinci (fig. 3, left). The force that must be exerted to slide a solid block on a table is simply proportional to the perpendicular force acting on this block: the more the block presses against its support, the greater the need to pull to make it slide [QUAKING BOWS]. The same principle applies to a capstan. By creating tension in the rope that is wrapped around the capstan (fig. 3, right), a perpendicular force is created that presses the rope against the cylinder. It becomes very difficult for the rope to slide. The more you wrap the rope around the capstan, the greater the frictional force. More precisely, it increases exponentially with the number of turns. Wrapping the rope three times around a capstan, a sailor can hold a rope under a literal ton of tension, while exerting a force equivalent to the weight of 1 kilogram.

A Self-Locking System

The same principle can be found in most knots, including the classic capstan hitch. Pull on one strand, and the outer loop strangles the lower loops to the stem. It is this pressure that prevents the rope from slipping but also keeps the tension in the knot. All knots secure their own strength using a comparable, self-locking system. No need to hold the free end, it will no longer slip.

The same type of effect occurs during spinning: the pressure at points of contact among the entangled fibers maintains the cohesion of the yarn and frictionally reduces the relative sliding of the fibers. This phenomenon is further magnified when the strand is pulled. On a small scale, the roughness of fibers (such as wool scales) or surface forces (so-called Van der Waals forces, or electrostatic interactions) also contribute to friction.

4 | *A capstan hitch: if you pull on the left strand, the right one being free, the outer loop crushes the inner loops, which prevents them both from slipping. This produces a self-locking system.*

EXPERIMENT

Here's how to dramatically illustrate the power of the capstan effect. All you need is a broom handle, a string about 1.5 meters long (roasting twine is perfect), a nonfragile mass of a few hundred grams (a small plastic bottle of water will do very well), and a much lighter object (a roll of tape, for example).

Attach the heavy and light weights securely to each end of the string (1). Check that the string is strong enough. Have a sturdy person hold the broomstick horizontally a string's length above the floor (2). The experiment consists in using the stick as a pulley to lift the heavier mass. Hold the light mass, then release it (3). If the experiment is well regulated, you will be amazed to see the heavy mass fall fast (4), and then stop suddenly (5): it will be miraculously held back by the string that spontaneously wound itself around the broomstick. Once a very small mass has wrapped itself around a stick several times, it's enough to hold back a big mass.

5 *After four turns around the capstan or winch, the rope that holds the taut sail is practically free at the other end. This is analogous to the unequal tensions between the full bottle and the roll of Scotch Tape in the experiment.*

Materials

1.

2.

Watch out not to knock over your sturdy friend.

3.

4.

5.

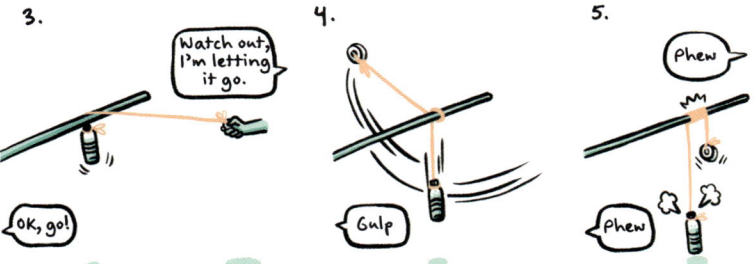

FOLDERS AND TAILORS: MASTERS OF VOLUME

Not recognizing themselves as the true mathematicians they are, clothes designers have a challenge: how does one endow the flat surface of a fabric with volume?

The ancient Greeks were masters of the art of draping, as evidenced by many sculptures of the classical age. That is when their technique reaches its climax, and their influence would be felt further during the Renaissance and neoclassicism. To drape oneself is to dress without having to adjust the fabric to the body. From loincloth to saris and capes, shawls, and ponchos, draped fabric is still a way of dressing in many countries. Beyond its elegance, it provides a simple solution to a

1 *Preserved at the Louvre, Diana of Gabii wears a short, flax tunic—a chiton—simply held by two belts and attached at the shoulder with a brooch. The subtle folds resulting from this assembly confer lightness and elegance to the tunic.*

geometrical problem that goes back to the dawn of humanity: how to use a surface—fabric—to cover a volume, in this case, the human body.

The Art of Folds

The most immediate solution? One method is to create permanent folds in the fabric, to create permanent drapery. The job of *creaser*, little known to the general public, requires competence in a craft the great fashion houses can't do without.

A creaser folds two identical sheets of cardboard in a periodic pattern, in the manner of origami. The simplest model is a series of accordion folds, though master creasers have more complex motifs in their catalogs (fig. 2): in Paris, Gérard Lognon, who has carried on this know-how, handed down from a long family of creasers since Napoleon III, now has several thousand models.

That's when the fabric steps into this dance: sandwiched in a mold consisting of two sheets of cardboard folded into a predefined pattern, the fabric is then compressed, moistened, and steam-dried for several hours, as it takes on that pattern of folds. This process, not unlike ironing, is doubly effective: it leads to robust and permanent folds; and the cardboard sheets are reusable for a new piece of fabric. This is how the folds of one of Empress Eugénie's garments can be replicated to embellish one of today's models!

> 2 | By squeezing the fabric between two carefully folded cardboard molds and then heating the whole, you get permanent pleats. Master creasers have thousands of different molds, rolled up like ancient scrolls on their shelves.

The Art of Cutting

There is a second technique for endowing fabric with volume: sewing pieces of fabric together. Prehistoric man used it to make clothes of animal skins sewn together with animal tendons. In Paleolithic times, the invention of the eye of a needle, which makes it possible to assemble two pieces with a single thread, was considerable progress for humanity at the time. Today, to produce fitted clothing, such as a bra or an epaulette, which cover complex, three-dimensional shapes of our bodies, a clothes maker might sew a dozen pieces together. But, again, why is it difficult to

3   Golf balls and potato chips are examples of surfaces with positive and negative Gaussian curvatures, respectively. It's impossible to perfectly apply a sheet of paper to them. The colored lines drawn on these surfaces signal curvatures with the same sign for the sphere and opposite ones for the potato chip. The product of these curvatures, the Gaussian curvature, is positive in the first case and negative in the second. It is zero for a cylinder whose surface contains a straight line, as with this can.

apply a fabric to a volume? And what geometric properties are at stake?

It should first be noted that this problem does not concern easily stretchable fabrics; with pantyhose or the body suits Olympic swimmers wear, stretchy synthetic fabric applies perfectly to the contours of the body. This obviously is impossible to achieve if the material doesn't stretch, such as a sheet of paper: one can easily affix a paper label to the cylindrical part of a wine bottle but not to the bulbous part of a bottle of Perrier.

Similarly, it's impossible to make a flat surface conform to the surface of a ball without stretching it. A geographer's planisphere "cheats" by modifying the relative distances of the spherical earth. The number of surfaces that can be created from a sheet are few: more than two centuries ago, the Swiss mathematician Leonhard Euler (1707–1783) listed these so-called *developable* surfaces. The simplest example is a cylinder.

The Tailor and the Mathematician

One must accept the fact that from our bulging shoulder to the crotch of our saddle-shaped pants, we are covered with non-developable surfaces, because of their *curvature*.

Curvature? We might understand the curve of a line, but a curved surface is more foreign to us. But it is not to the great mathematician Carl Gauss (1777–1855). Fifty years after Euler, Gauss generalized his work on shapes thanks to the notion of Gaussian curvature: a positive curvature evokes a sphere; a negative curvature the saddle of a horse; and zero curvature, such as a cylinder, necessarily contains a straight line (fig. 3).

Gauss showed that this curvature is preserved when a surface is deformed (without modifying its dimensions): a flat paper surface therefore can never be transformed into a horse saddle or be made to bulge.

That is the impossibility clothing designers confront facing the shoulder of a shirt. But how does assembling several pieces of cloth then make it possible to get a doubly curved surface?

Let's play tailor's apprentice: starting from a disk from which we've removed a section before picking up its lipped edges, we get a cone that can't be flattened without creating folds (fig. 4). By cutting and gluing, we shorten some lengths, and thus circumvent Gauss's restriction. This yields a surface that is no longer developable and starts to look like a dome.

In fact, we can't speak of curvatures with respect to these structures made with scissors because there are *singular points*, such as the top of the cone where the surface is not smooth but angular. To complete Euler's and Gauss's studies, the French mathematician Henri Lebesgue (1875–1941) introduced these pointed shapes, which are also manifest in crumpled paper [FOLDING AND CRUMPLING PAPER BALLS].

One can also reverse the operation, that is to say insert an additional angular section into a notch made in the paper disk. We then get a wavy surface reminiscent of a horse saddle. This principle of inserting extra material is used in gored skirts: the cylindrical shape of an otherwise straight skirt flares, thereby enhancing the silhouette. Thus, couturiers, who understand the importance of the geometry of their patterns, apply advanced mathematics to dress our volumes elegantly using simple pieces of cloth.

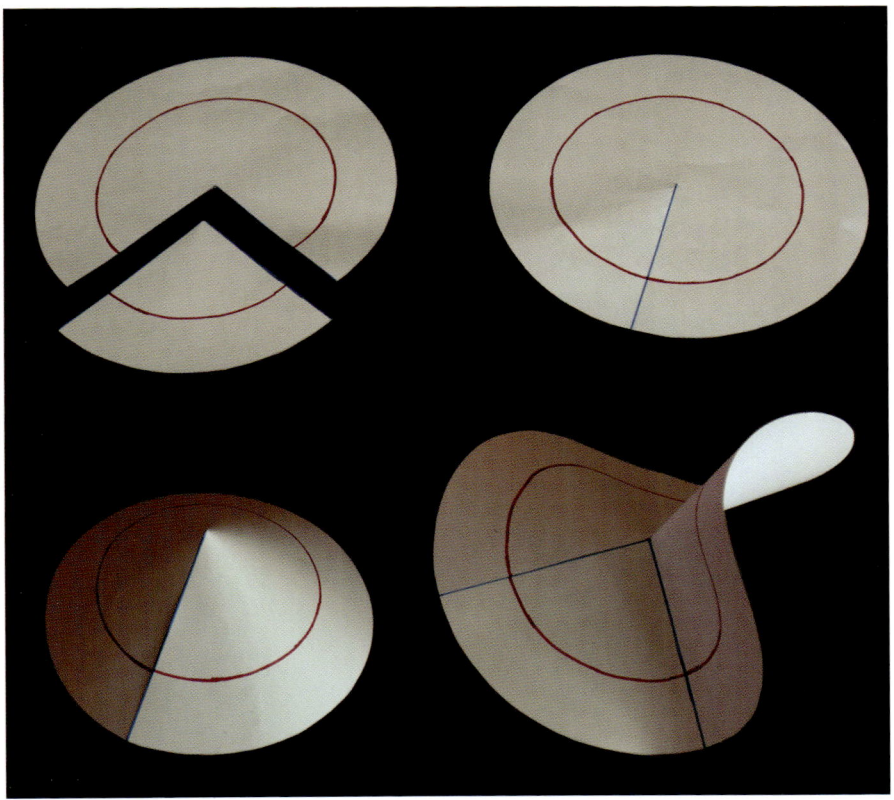

4

A rigid paper disk (upper left) from which an angular section has been removed adopts a closed cone shape (bottom left). Take this same section, and now insert it into a new disk after having cut a notch (blue line at the upper right): the resulting conical shape is wavy. These surfaces, obtained by cutting and gluing, are no longer developable, since it is impossible to flatten them.

EXPERIMENT

Take a look at a classic, traditional soccer ball: proof that one can roughly cover a spherical surface from cuts and seams. It consists of 12 pentagons and 20 hexagons.

A more contemporary soccer ball, the *Brazuca*, created for the World Cup in Rio, seems to defy geometers. It uses six cross-shaped sheets, centered at the vertices of an octahedron and then assembled. You can reconstruct them from the following pattern, which you will have to patiently trace and cut into six pieces and then assemble using the tabs. If you like challenges, you can also build a closed ball with only two sheets sewn together continuously: you'll come up with a baseball.

5 | *The* Brazuca *soccer ball, using six identical pieces, created for the 2014 World Cup in Rio de Janeiro.*

WEAVING AND BRAIDING

The unique, mechanical properties of fabrics, related to the entanglement of threads, make it possible to challenge the laws of geometry. How do we use fabric to closely cover the curvature of our bodies?

At first sight, seamstress Madeleine Vionnet made a dress in 1920 that evokes the one worn by the famous statue Diana of Gabii [FOLDERS AND TAILORS: MASTERS OF VOLUME]. And yet, the principle of its making is quite different and connected very precisely to the fabric's *bias*, as we will see.

1	*Dressmaker Madeleine Vionnet's handkerchief dress was created using a number of minimalist stitches and four pieces of square fabric. It shares purity and simplicity with the tunics of antiquity, while bringing into play an original construction principle that makes use of the fabric's bias.*

Creating Fabric

Let's clarify this and see how fabric is created. If you have ever visited a *souk*, an Arab marketplace, you may have seen a loom in action. One of the most beautiful historical looms, from 1801, is preserved in Paris at the Conservatoire National des Arts et Métiers and bears the name of its inventor Joseph-Marie Jacquard. Another working Jacquard loom can be seen at the Shelburne Museum in Burlington, Vermont. The working principle of these looms is still relevant. It is also special in that it uses punch cards to control the play of threads that weave the fabric; this programming system makes it a true ancestor of the computer.

On a loom, the warp threads are arranged in parallel, along the entire length of the machine. The weft threads cross them at right angles, alternately running over and under a warp thread. The stretched, overlapping threads exert pressure at their points of contact, which guarantees that the fabric holds its shape [AN IMPRESSIVE BRIDGE OF GRASSES]. In *plain weave fabrics*, each weft thread alternates, above and below the warp threads: the top side and its flip side appear identical. There are other possible crossovers: in a *twill*, weft threads skip one warp thread. In the case of *satin*, more threads are skipped; with less dense overlapping, weaving provides more give, more freedom to deform (fig. 3).

A Victory at the America's Cup

A fabric has its particular mechanical properties. Stretching it in the direction of the warp or weft (*following the grain*) is difficult, but easier along its diagonal line (or *bias*). The angle of the crossing of the two types of threads varies during the pull: the crossings transform the small square meshes into diamonds. The contraction one detects of the diagonal opposite the stretching is a result of this geometrical transformation, which keeps the overlapping points steady (fig. 4).

It is said that one of the first America's Cup races was won by an American crew who took the following principle into account: for a sail not to

2   A Jacquard loom. The warp threads stretched from left to right are alternately lifted to pass the weft threads through at right angles. Perforated plates that scroll and drive the lifting of the frame are at the top of the structure.

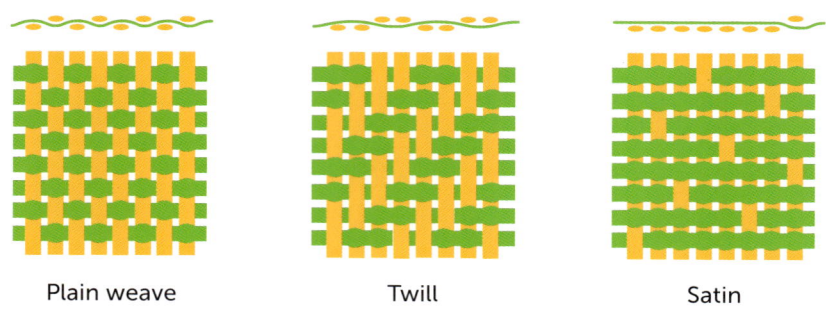

Plain weave   Twill   Satin

> 3 — *From canvas to satin, the density of warp and weft overlapping diminishes. Decreasing the number of crossing points makes the textile smoother and more deformable.*

deform at the cost of losing some wind, it must be cut so that it is aligned with the sail's edge, which was not the case of British sails cut on the bias. At the finish line, Queen Victoria attended the arrival of the American sailboat and asked where the English ships were; she is said to have been told "Ah! Your Majesty, there is no second!" The English sails were adjusted for the very next competition.

In the 1920s, Madeleine Vionnet understood that cutting fabrics at a right angle enabled deformations that straight lines proscribed. She would take advantage of the fact that a dress made of four handkerchiefs, held vertically on the bias, hangs vertically under the effect of its own weight and slims down in a transversal direction. As a result, such a dress naturally adjusts to the waist of its wearer. A new chapter opened for haute couture. Pamela Golbin, who remained in charge of Vionnet's collection, points out that thanks to this cut an ancestor of ready-to-wear clothing had been created. The virtuosity of the great lady's creations earned her the moniker of the *Euclid of fashion*.

Realizing the Impossible

We have seen that it is impossible to apply a sheet of paper to a sphere without tearing it or creasing it [FOLDERS AND TAILORS: MASTERS OF VOLUME]. Mercator's sixteenth-century maps are hugely stretched at the poles and reduce the apparent surface of Africa. Tailors are a bit luckier, as the Russian mathematician Pafnuty Chebyshev (1821–1894) demonstrated. At a congress in Paris in 1878, he published an article titled "On Patterns in Making Clothes" that asked the following question: can one dress a sphere from basic cloth squares? Chebyshev offered a positive answer: it is possible, at least locally, by using the available deformations on the bias, as shown by applying fabric to a balloon. The solution is not merely academic. It is now possible to calculate and produce fabrics using *irregular* mesh sizes, in order to get them to take on a desired shape later without having to stretch the piece. Since we know how to reproduce the external shape of a human body computationally, making a fitted item of clothing using a single, seamless piece is a priori no longer so difficult.

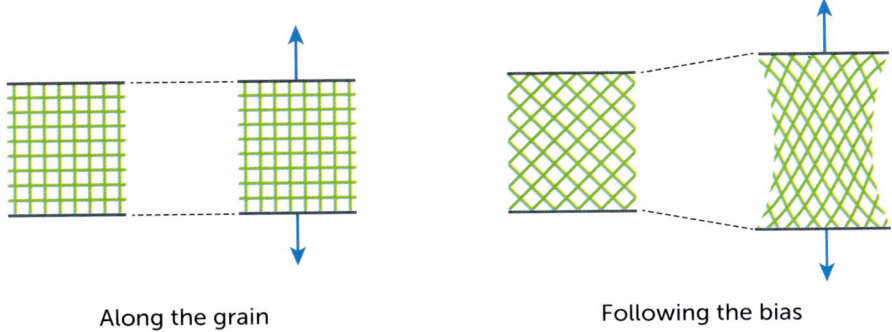

Along the grain   Following the bias

4 | *On the left, a fabric pulled along the straight grain is slightly deformed. On the right, pulled on the bias, it stretches in one direction and contracts in the other.*

Braiding

On a bigger scale, weaving plant-stem structures, in particular wicker, is not unlike weaving yarns. The thickest stems are pre-soaked to soften them, and then braided. They make up a frame that supports the finer fibers, arranged later according to the same principle as fabrics. This construction also evokes how certain nests are woven [BIRDS AS ARCHITECTS].

Eating Wild Cassava

Weaving natural fibers, whether for deep or flat baskets, partitions, or fly swatters, illustrates the convergence of ancestral techniques developed by peoples of independent cultures on all continents. The Amazon's *sebucán* is a remarkable example of how braiding vegetable matter can produce both beauty and function at once. It is also called a "snake" because of its approximately 2-meter-long tubular shape that is closed at its base. Amazon people use it to filter freshly grated wild cassava powder, which contains a toxic juice. A lever system forcefully stretches the *sebucán* once it is filled with cassava. Because of the braiding, the tube narrows, thus compressing the powder and exuding the juice.

This mechanical action thus produces the same effect of simultaneous narrowing and lengthening that Madeleine Vionnet observed in fabric. In contemporary technology, the principle of this deformation also became a source of inspiration in robotics. The challenge was to power artificial muscles by inflating them with compressed air and then deflating cylindrical bladders equipped with a woven structure. As with a *sebucán* that shortens as it is filled, pressurization causes artificial muscles to contract.

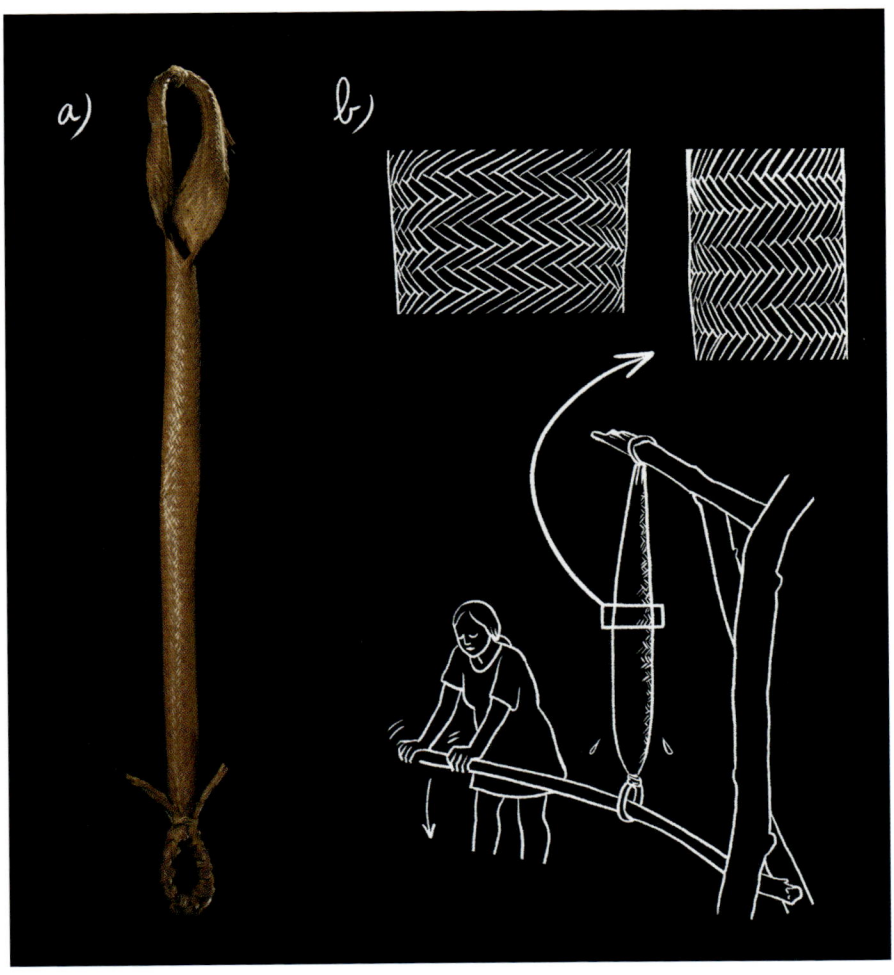

(a) A sebucán at rest, before being filled with cassava. (b) A comparison of the same segment of the interweaving of a sebucán, at rest (left), and stretched (right)—and the scale of the two images is identical. The stretched shape is smaller in volume (it is proportional to the radius of the sebucán).

EXPERIMENT

Perhaps you've played around with those two-colored braided tubes, called Chinese finger traps, into which you stick a finger at each end—and then can't free your fingers. The tube narrows as soon as you try to remove your fingers by pulling at the ends. This structure is also used in medical gauzes that are slipped on a finger and squeeze the finger when pulled.

Make a Chinese finger trap using four identical color strips that cross at right angles, one after the other. Then braid it regularly by alternating the colored strips (2). To set it up, use a cylinder to serve as a matrix, to which you attach the four bands beforehand (1).

> **6** *By pulling on the structure after you've inserted your two index fingers without any trouble, the tube's diameter decreases as the tube itself gets longer: and your fingers remain trapped!*

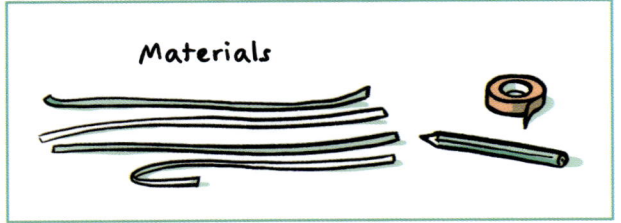

Materials

1. Set-up

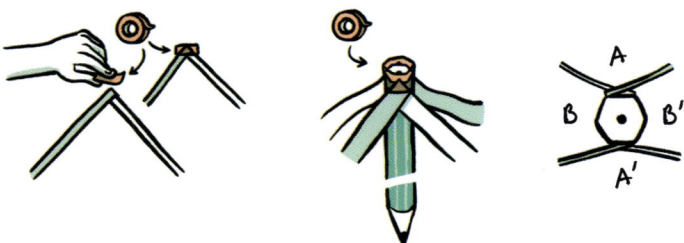

A

B B'

A'

2. Braiding

Side B & B'   Side A & A'

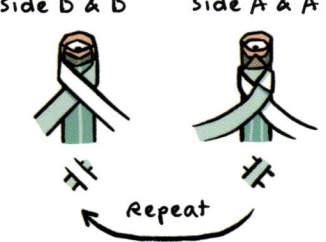

Repeat

3. Boom

FOLDING AND CRUMPLING PAPER BALLS

Thanks to how easily paper changes shape, sheets of paper have become elements of artistic creation. They benefit from the disorder of crumpling or, conversely, from the systematic organization of their folds. Moving from a planar surface to a three-dimensional object can be a matter of folding.

Who would have thought that a paper ball would be the site of exciting phenomena? In reality, the mysteries of a paper ball reside in folds. Most of us only become interested in this when we remove a crumpled piece of clothing from our closet. But just think of elegant origami sculptures or the work of painter Simon Hantaï (1922–2008): folds are hardly a curse,

1 | *Étude, Simon Hantai, 1969. The randomly folded canvas is covered in blue paint. Once opened, it brings out in white the irregular patterns of the folds.*

quite to the contrary! The artist covers crumpled canvas with color then stretches it out. The intricate network of folded areas appears hollow and immaculate, evoking some superb abstract composition.

The Irremediable Marks of Crumpling

The subject that underlies these aesthetic considerations is, of course, the physics of the fold, whose mathematics we learn about when we deal with cloth or fabrics [FOLDERS AND TAILORS: MASTERS OF VOLUME]. Here is an epistolary example: to slide a large sheet of paper into a small envelope, you must fold it sharply and strongly so that the folds remain. But perhaps you have started your letter several times before sending it, without noting this seemingly contradictory fact: that when you started to crumple the drafts, marked folds appeared very quickly without having been creased on purpose. Why is the sheet damaged as soon as you start making it into a ball?

The answer is simply that when you've bent a sheet in one direction it is very difficult to bend it in another. Try the experiment by bending a sheet slightly in the shape of an arch: now try to bend it perpendicularly to the curvature. You will not succeed without damaging it. And if you persist in your efforts, crumples will eventually take shape, delineating almost flat areas on the sheet. In the end, your completely crumpled sheet of paper will look like a patchwork of small areas, each curving its own way. The geometrical difficulties the sheet encounters materialize and are concentrated on these reduced areas, namely at the folds.

In passing, let's note another oddity: if you uncrumple the ball, the crumpled sheet of paper bends less under its own weight than did the initial smooth sheet. Indeed, all local, randomly oriented curvatures that we've described (those in small areas) increase the resistance to curvature in all directions. In contrast, a crumpled sheet is easy to stretch out: the excess length stored by the folds is easily released.

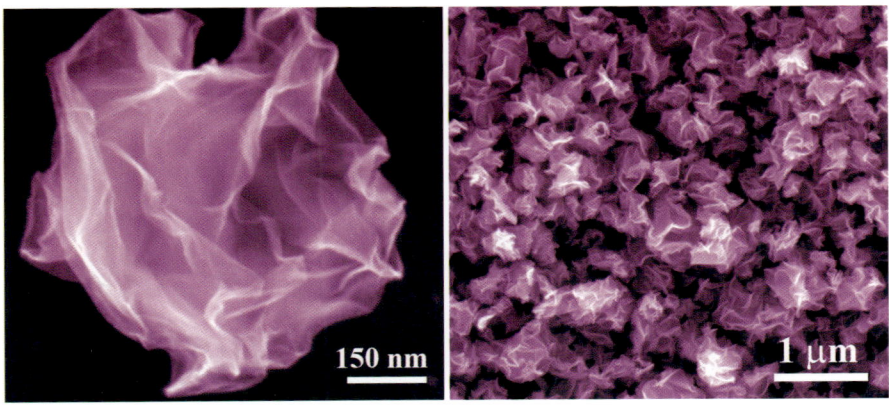

2 | *An isolated crumpled graphene dumpling (left). A set of small paper pellets stacked against each other (right). Having an unrivaled surface-to-volume ratio, these pellets probably will be used as catalysts in tomorrow's fuel cells.*

Crumpling Graphene

The remarkable mechanical properties of crumpling intrigue many physicists. Some have even hitched themselves to studying the thinnest sheets in existence: graphene sheets. It must be said that they consist of a single layer of atoms, in this case of carbon atoms that are one ten-billionth of a meter thick—a million times thinner than a hair! For their thickness, these sheets are mechanically stronger than our best traditional materials. The challenge now is to combine the electronic and mechanical properties of graphene in designing new components and sensors, the ones we may soon find in our smartphones.

Take these ultrathin sheets that, when put in a solution that is then allowed to evaporate, shrivel into pellets under the effect of capillary forces. Even if they look a lot like their big paper cousins, this unusual material offers totally new properties that are just beginning to be explored. Put side by side, these pellets, stiffened because of their folds, form continuous layers. Used as material for electrodes, these layers make it possible to attain capacities, given equal mass, far beyond any other material used today in lithium batteries—and that is thanks to a large conductive area as well as a broad permeability and affinity to metal ions.

Mastering Folds

To understand and control crumpling, physicists began by isolating one of these points that appears in a crumpled sheet at the meeting of several folds. How? A simple set-up is to lay a sheet flat on a drinking glass and press the tip of a pencil down in the middle of the sheet. This shapes a particular sort of cone, since it produces a ripple on the side and does not conform all the way around the edge of the glass (fig. 3).

At the top of this cone, the sheet is very deformed locally and irreparably damaged: the slight contact of the pencil has seriously damaged the material by effecting a geometric concentration. But sacrificing a point is obviously less costly than stretching the sheet over a larger area.

On a crumpled sheet, these singular points are numerous and are connected by folds that act as many hinges enabling us to refold the sheet. The rigidity of these hinges, however, limits the compactness of the final pellet. The random organization of the folds, moreover, leads to many points of contact and empty spaces. This is why carefully folded clothes take up less space in a suitcase than if they are carelessly and randomly crammed in.

The Mathematics of Origami

Origami takes advantage of these very properties in the careful arrangement of folds to create real paper sculptures. Recently, mathematicians have even developed some curious algorithms: they determine the geometry of folds to create any three-dimensional shape desired, such as that of a rabbit (fig. 4). These techniques also interest engineers. Like sailors who meticulously fold their sails to be able to store them in ships' vaults, or parachutists who would never delegate the work of preparing

3 | *A flexible sheet that one tries to apply continuously to the edge of the glass necessarily creates a fold that ends in a point. Another solution is to remove a section and re-glue, as presented on p. 129.*

their parachutes, space engineers are required to fold up panels during space flights. Inspired by examples in nature, such as hornbeam leaves that are about to bloom, Japanese astrophysicist Koryo Miura invented a very compact, periodic fold that has occasionally been used for solar panels.

Origami and crumpling are thus kindred expressions characterized by the same geometrical constraints; and some artists, such as Vincent Floderer, do not hesitate to wed them, associating order and disorder to form complex and elegant structures, inspired by shapes of living things (fig. 5).

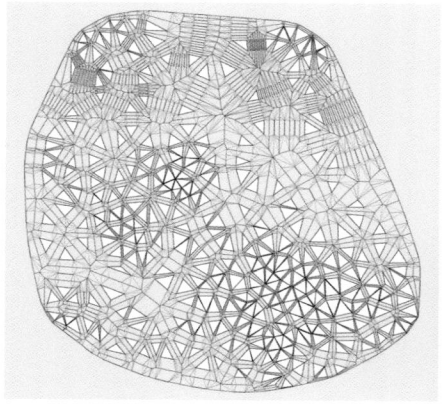

4   *It is very difficult to discern what pattern an origami generator creates automatically (left); here, the shape resulting from the geometry of the folds is a rabbit!*

5 Frou-Frou *by Vincent Floderer. This is very fine paper strongly creased around several precisely selected points. When reopened, these organic, asymmetrical, and vaporous shapes combine order and disorder aesthetically.*

EXPERIMENT

Let's go further. Make a giant paper ball with a double newspaper page. Then do it with a single page, and a half page, and a quarter page, and so on, making paper balls the same way. Now evaluate their average diameter. Does the size of balls evolve like that of a compact mass? In this case, the mass of the balls should increase by the cube of their radii. In other words, multiplying the initial mass of newsprint (or, equivalently, its planar area) by 8 means the radius should be multiplied by 2.

In fact, a big crumpled ball is less compact than a small ball, so that the radius increases faster than expected. It is to account for this surprising observation that such balls are classified as fractal objects, characterized by an intermediate dimension between that of a surface (2) and that of a volume (3). For math fans, a Mexican team estimated the fractal dimension of crumpled paper balls to be 2.27 ± 0.05. We obtained a value of 2.45 ± 0.05 with a free daily newspaper. And what about you, how much do you come up with?

Would you like a small challenge? How many times do you think it is (humanly) possible to fold a sheet of paper? Try; you will have a hard time exceeding seven or eight successive folds. It seems that the world record, with a very long strip of thin paper, was set by folding it on itself a dozen times!

Materials

CRUUUUUNCH...

The most relaxing experiment in this book.

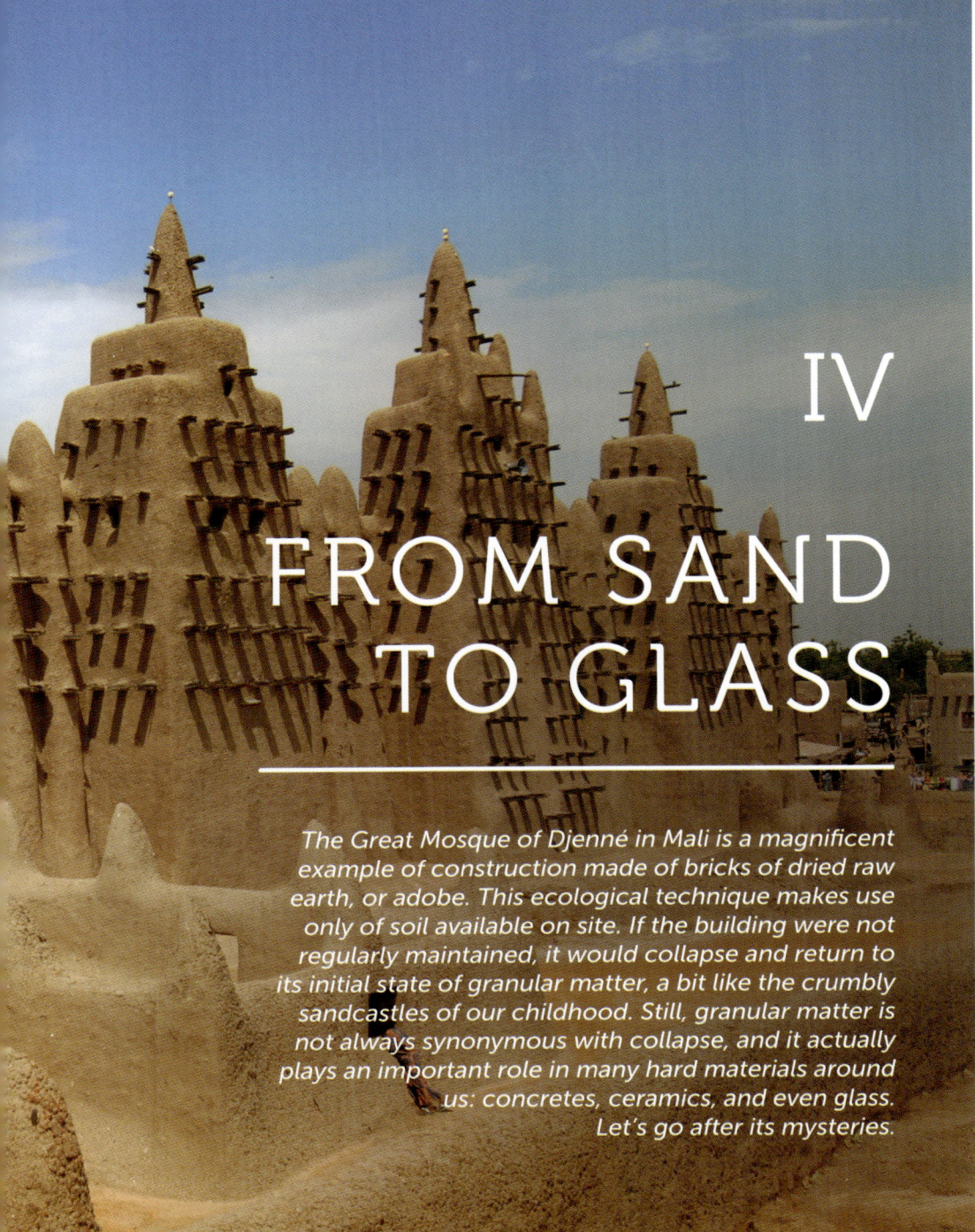

IV

FROM SAND TO GLASS

The Great Mosque of Djenné in Mali is a magnificent example of construction made of bricks of dried raw earth, or adobe. This ecological technique makes use only of soil available on site. If the building were not regularly maintained, it would collapse and return to its initial state of granular matter, a bit like the crumbly sandcastles of our childhood. Still, granular matter is not always synonymous with collapse, and it actually plays an important role in many hard materials around us: concretes, ceramics, and even glass. Let's go after its mysteries.

SEEING A WORLD IN A GRAIN OF SAND

All you need to do is to take a walk on a beach to recall that the surface of our planet is largely covered with grains of sand. How can we learn about the varied origins of this sand? What was their path? Let's get a magnifying glass and investigate!

Surprising Grains of Sand

In California, Glass Beach is a curious testament to nature's ability to compensate for being neglected by human beings. This site was once a sad, open-air dump where car bodies, appliances, broken glass, dishes, and more piled up until the authorities decided to clean it up in the

1	*The astonishing Glass Beach in California. These grains of vitreous sand attest to the past of the site: a sort of dump that has aged beautifully.*

1960s. Scrap merchants salvaged the metal parts while waves made short work of dangerous glass and ceramic shards by breaking them into small pieces and gradually blunting them. A few decades later, this erosion has yielded multicolored sand that is popular among tourists.

In the wild, shores and deserts are not the only places where granular materials accumulate. All you need to do is walk along a river or through a plowed field to notice the huge variety and ubiquity of these sort of atoms of landscape: after air and water, granular materials are the most prevalent substance on Earth. They have a long story to tell. The poet William Blake (1757–1827) aspired "To see a World in a Grain of Sand ... / Hold Infinity in the palm of your hand." Whether dust, rock, or sediment at first, individual grains in fact humbly witness the processes that shaped them.

Traveling Sand

Grab a handful of sand and let your mind wander for a moment. The sand's texture is already rich in information. If it flows between your fingers, that's because the grains have smoothed out, having rolled around in a riverbed or in the ebb and flow of beach waves. Other grains of sand are at first rough and angular and then are shaped by crashing about.

The same handful of sand is made up of grains that reveal their varied geographical origins through their appearance. Want a few examples? Some of the translucent grains are made of silica: they may come from the degradation of sandstone—a rock that was itself shaped by the slow aggregation of solid particles sedimented at the bottom of a sea. Grains of sand can also come from degrading granite rock that manifests in the several sparkling colors of feldspar, quartz, and mica. Tahiti's black sand beaches recall their volcanic origins—and then there's the sometimes painful reminder for tourists that accumulated sunlight causes its fine-grained basalt to heat up and burn their feet.

And what about white sand? It often comes from the fragments of skeletons or the shells of living beings (shellfish, corals, foraminifers). Sometimes the entire structure of their mineral envelope reaches us intact. A grain of sand becomes a source of wonder, like the star-shaped sand of the beaches of Okinawa, Japan (fig. 2)! According to local legend, these tiny grains are the fruit of a cosmic union between the Polar Star and the Southern Cross. A monstrous marine snake is said to have attacked it, leaving behind only beached skeletons, enough to humble us by reminding us that foraminifers—or the human beings that we are—are mere "stardust."

2   *Sample of white sand collected at a beach in Okinawa, Japan. These star-shaped grains, measured in millimeters, are made up only of foraminiferal mineral shells. This is all that remains of the remarkable marine organisms equipped with tiny arms that enable them to move and feed.*

3
The horizontal layers of soil—its horizons—tell the local history of the subsoil. It begins with the very thin, brown surface layer, which is rich in humus and bears life. This clay-containing soil forms the basis of adobe architecture [EARTHEN ARCHITECTURE]. The successive and very different types of layers consist of matter that comes from earlier rock erosion.

Earth's Granular Soil

Other granular materials have an even greater variety of textures and colors than the beaches of the world's oceans: namely, the components of the surface layers of the earth (fig. 3). In this case, we can admire the play of shapes and colors of the often parallel layers, or *horizons*. If we took a short trip to the center of the earth, what would we see? First, the *soil*—a thin layer of fertile humus where plants take root in a few decimeters to a few meters of earth depending on where we are on the globe. Stemming from the very slow degradation—sometimes hundreds of centuries—of the bedrock, this layer is home to an unsuspected underground life where myriads of bacteria, plants, fungi, insects, worms, moles, and more, thrive in a common ecosystem.

Digging a little deeper, we would discover a subsoil sometimes composed of layers of clay or sediment of much finer granulometric measurements. Deeper still, the subsoil leads to the bedrock whose nature changes with the terrain. The sediments in the subsoils of Paris remind us, for example, that these basins were once under water. In contrast, basalt found in the Columbia plate in the western United States comes from slow lava flows that lasted millions of years.

Artificial Granular Material

We build buildings from "small" to "big": it takes billions of grains to build a sandcastle [THE SECRET OF SANDCASTLES]. But our natural resources in granular matter seem more limited today, and sand is not an easily renewable resource. We are therefore called upon to laboriously recreate the patient work of nature: to produce "small" materials from massive ones. This is the work quarry crushers do: reduce the size of rocks to gravel and sand. Or of cement mills, those long, slightly tilted barrels that break down hard cement *clinker*, itself the product of an earlier firing of a mixture of limestone grains and silica. In the same way, turning wheat grain into flour is done in impressive flour mills. These crushing processes, however, are not very efficient: on average we grind a ton of material each year per planetary inhabitant. Friction causes most of the energy supplied to break up solids to be lost as heat: simple milling accounts for the largest amount of energy consumed on Earth. The challenge to engineers is to find more ecological solutions.

EXPERIMENT

How does one sort the granular material in a clump of soil by size? You could sift the grains in a series of strainers or sieves with increasingly fine openings, each time setting aside grains that are larger than the size of the mesh. (In figure 4, we used sieves with 20 and then 10, 5, 2, and 1 millimeter openings): the coarsest level is gravel, then sand, and finally clay soil.

Another approach, which gold diggers used, is sedimentation. Pour a few inches of soil into an empty jar and fill it with water (1). Close or cork this suspended mix, shake it, and let it rest (2 and 3). The larger grains settle quickly to the bottom. Progressive layers, of finer and finer granulometric size, continue to pile up on top of them. As for the very fine particles, they remain suspended almost indefinitely and muddy the water. You can observe a similar effect when you go swimming in a lake's crystal-clear water. The whirlpools your movements cause stir up and disperse the sediment's fine particles: and now you've muddied the waters!

4 *By sifting some rammed earth, we can make note of the proportion and size of the various grains that it consists of: from pebbles to granular materials, gravel, sand, and clay from ancient glacial moraines.*

Materials

1.

2.

3.

4.

THE SECRET OF SANDCASTLES

Now that you think about it, isn't it amazing that we can build with sand? The strength of a sandcastle illustrates the effect of the capillary forces that come into play between the grains of sand and water. Or, how a liquid behaves like cement.

I sing the little war
Of the brave children of old
Who at the beach battled
To save a sandcastle
And its unbreachable ramparts
That a wave would sweep away.
  —Georges Brassens, "Castles in the Sand"

> 1 | *This sand art was created on a beach in Copacabana. Such a finely chiseled structure can stand thanks to the forces of attraction at play between water and the grains of sand.*

A bucket with a crenellated bottom for towers, a wall, moats … and presto! We've recreated the sandcastle of our childhood. Our achievements remain quite crude compared to the masterpieces created by today's sand artists (not burrowing creatures but people, arenophiles, who "love sand") at the beach in Copacabana, Brazil, and elsewhere. Their marvelous sculptures, with fine arches and overhanging motifs, seem to defy the laws of gravity.

Faced with these ephemeral structures, a physicist might wonder: how are they able to stand? Thanks to the water, of course, as every child knows. Sweet, salty? Whatever. Can water be replaced with oil or another liquid? Why not?—the castle would keep its integrity! Need a little bit or a lot of water? Enough, but not too much—the sand must be damp. But how is it that these forces turn a liquid into cement?

Like Attracts Like

In a very general sense, identical molecules attract each other. It is thanks to this property that matter is not all gas but exists in condensed forms—whether liquid or solid. In the first century BCE, the philosopher Lucretius imagined particles of matter as equipped with small, more or less sturdy hooks. In this view of things, materials were like fabrics that "hooked up." It would take two thousand years and twentieth-century scientific innovations to, in a way, validate this great intuition.

One can imagine molecules at the surface of a liquid or a solid as brandishing handles pointing externally, outside of the interface, hoping to hook up with another molecule for which they have affinities. Obviously, in fact, intermolecular forces are at work at a distance.

Logically, two grains of dry sand should adhere at their point of contact. Admittedly, this adhesion does exist, but the contact area is extremely small because grains of sand are solid and are unable to deform enough to conform to the topography of their neighbors. In practice, as soon as the size of the grains is greater than that of a speck of dust (typically a few micrometers), gravity is enough to defeat such frail interactions.

A Simple Drop of Water Creates a Bridge

A drop of water partially lifts this obstacle by building a bridge between two grains of sand (fig. 2). In fact, the liquid is able to deform and conform to all the crevices on the surface of the grains, thus considerably increasing the contact areas. The resulting attraction is called a *capillary* force. The same force tends to make our hair stick together when we wet it [WET HAIR]. Its intensity is modest but sufficient for the grains of a sandcastle a few dozen centimeters high to withstand the weight of the structure.

The force exerted by a liquid bridge between two grains is related to the curvature of its wall, that is to say, to the interface between water and air. If, as here, the interface is convex, the pressure inside the liquid is lower than the atmospheric pressure. This results in suction that causes the grains to attract one another.

2   *Wetting two marbles makes it easy to see the water bridge that is established between them, just like in a sandcastle. But, although this bridge guarantees cohesion in a structure on a beach, it is too fragile to actually support the weight of a marble.*

The evolution of capillary forces with the volume of liquid bridges that are created is surprising: capillary force grows very rapidly as this volume increases, reaches a plateau, and then gradually decreases until it vanishes when the grains are completely immersed. In other words: without an interface, there is no longer any capillary force. At its optimum, the force of cohesion is proportional to the radius of the grains, while their weight varies as the cube of the size. Conclusion: with coarse grains, gravity prevails rather quickly. Attempting to build castles with coarse gravel is therefore doomed to failure.

Castles with Clumps

Creating a sandcastle ultimately requires careful measuring. The sand you select must be fine and just moist enough to establish a maximum number of capillary bridges. But be careful not to wet it too much because that tends to bring together these bridges and diminish their effectiveness: a drowned castle will collapse. Besides, have you ever noticed how at low tide, far away from the water's edge, sand is dry and loose, then seems to harden as you get nearer the water, and finally becomes soft at the edge of the water?

By its very presence, water thus acts on the strength of a granular material. This is why humidity is an essential factor in civil engineering if one is interested in the compactness of wet soils and their cohesion. It is this precise measuring that sandcastle builders seek to master, just as public works engineers do. And what if water did not wet grains of sand? That's the crazy idea that toy manufacturers came up with: they developed a "magical" hydrophobic sand that stays dry when immersed in water. Even if it is utopian to build with this kind of sand, it is nonetheless perfectly possible to create ghostly underwater structures! Submerged grains are surrounded by an air pocket that ensures the cohesion of the

> | 3 | A classic block of sand (left). Adding water in large quantities inevitably causes it to collapse. "Magical" sand made of hydrophobic grains (right) makes it possible to create a totally submerged block. Unlike its beach counterpart, it disintegrates as soon as it is out of the water.

whole. This effect is a major constraint for manufacturers of chocolate powder, as the powder tends to get lumpy when milk is added for the same reason. Researchers are developing strategies to disintegrate these underwater grain structures.

Soils and Concretes

Unlike beach sand, soils and concretes are particularly hard when dry. Do these granular materials have something special or extra? Granular materials kneaded in a concrete mixer are very similar to wet sand. Masons, however, have to add a major ingredient: cement—very fine particles that act like a rigid glue after setting; in the case of soil, they are clay particles. Nevertheless, the physics of sandcastles helps us predict the behavior of these materials as they are processed [LIQUID STONE: CONCRETE]. The art of sandcastle builders is therefore essential to concrete manufacturers and soil specialists.

4 | *A pile of sand on a dry (left) or wet (right) surface under the same conditions. Water absorption by capillarity gives the sand the necessary cohesion to prevent avalanches and lead to the formation of slender columns.*

EXPERIMENT

Using a funnel filled with sand, form a jet that pours into a saucer. The sand falls in a pile—a cone, of the most ordinary sort—as in repeat avalanches (1).

Now, after pouring a little water in the saucer, repeat the experiment again, (2). By joining the already formed pile, the dry sand quickly absorbs the liquid, which reinforces its cohesion and prevents collapse. You'll be able to see, alternately, the erection of high columns of wet sand (3) and their collapse when they are too slender (fig. 4).

How high can you erect these totem poles of real sand? How does this change with the size of the granular material?

Materials

1.

2.

3.

EARTHEN ARCHITECTURE

Earthen constructions have exceptional qualities that many animal species and more than a quarter of humanity have noticed. What are the secrets of their strength and durability? Are they a solution for sustainable development?

The ziggurats of ancient Mesopotamia, the pyramids of El Fayoum, Morocco's ksars, Mali's and Burkina Faso's wood-scaffolded mosques: these buildings are striking for their extraordinary architecture while they offer a deep sense of harmony. How does this happen? They are built of sun-dried earth. These earthen habitats, often integrated into rural landscapes, use raw materials from the available resources in their immediate vicinity.

Such habitats are durable as well as strong. Did you know that the oldest skyscrapers in the world, in Shibam, Yemen, are made of earth and

1 | *These earthen skyscrapers in the Yemeni city of Shibam are the oldest in the world.*

date back to the sixteenth century? Or that part of the Great Wall, which extends more than six thousand kilometers in northern China, consists mainly of material made from layers of plant fiber and earth? This composite made it possible to use local materials while sparing water resources, which are generally scarce in these areas. Although partially damaged since it was built several centuries before our era, the Great Wall remains the largest construction ever undertaken—and of such admirable robustness.

Eight-Story Buildings

A very long history precedes this type of construction. Many earthen buildings, and even entire neighborhoods of them, are listed as World Heritage sites. These are often places of worship or places performing a social role such as forums or markets—that is probably why they have survived through the ages. The ancient part of Shibam City, known as the "Manhattan of the Desert" owing to its mud buildings rising up to eight stories and 30 meters high, is one such site (fig. 1).

Despite the fact that wind and rain slowly erode all things in this world, wars remain the real enemies. The bombings of Sanaa's skyscrapers in Yemen and of the famous Mostar Bridge built in Bosnia in the sixteenth century under Suleiman the Magnificent are sad examples of destructive human madness. Fortunately, the Mostar Bridge was rebuilt similarly, using materials of the same type as in the original structure: in particular, a mortar was identified that contained horsehair and yolk to improve adhesion among the various elements.

Building Using Earth

No need to rush to exotic lands to admire raw earth habitats, as they are readily available to see. It is estimated that a quarter of the world's population currently occupies housing of this type—and it represents about 15 percent of France's architectural patrimony, even though it is partially

2 · *Built after the earthquake that struck China in 2008, this earthen construction received a Terra Award. It is inspired by traditional community residences,* tulous, *built by the Hakka people of Fujian province, China, ever since the fifteenth century CE.*

hidden behind a rendering layer. Several construction techniques have been developed that take advantage of local resources. *Rammed earth* is made of earth containing grains of various sizes [SEEING THE WORLD IN A GRAIN OF SAND] and clay. This mixture compacted in formworks is a true, natural concrete. *Adobe* consists of bricks that must be molded and allowed to dry before being assembled. *Cob* consists of clay and vegetable fibers: applied on wooden frames, it is at the heart of the *half-timbered* structures in our old neighborhoods.

There's nothing nostalgic about these techniques: current research has inspired many architects, and earthen construction today seems like a solution for humanity. Organized by NGOs with UNESCO's acknowledgment, the Terra Award recognizes contemporary architects who prioritize the value of raw earth. A recent example of this is the reconstruction of the Chinese village of Ma'anqiao (fig. 2). It is reminiscent of the collective habitats in the south of the country, the *tulous* (literally "earthen buildings"), which the Chinese have built for five centuries and which served as community residences. The benefit is as aesthetic as it is economic and ecological.

Choosing the Right Granular Material

But what ingredients keep these earthen structures in place? First of all, a granular medium whose large granulometry, from pebbles to fine sand, must be precisely controlled. Including small grains in the mixture, for example, helps reduce these structures' *porosity* to water as well as the capillary absorption of moisture that comes from the soil.

Adding fibers has the advantage of consolidating the structure. While animal horsehair was used in the Mostar Bridge's mortar, hemp and straw are much more common reinforcing materials. Beyond gaining more mechanical strength, these fibers also play an important role in thermally regulating habitats.

Another constituent is water. In an apparently dry earth wall, there remains a small percentage of liquid water trapped in the form of capillary bridges between the clays. This water acts like an adhesive allowing the cohesion of the granular medium [THE SECRET OF SANDCASTLES]. This is not, however, about building sandcastles that collapse once they have dried! Keeping earthen structures sustainable would be impossible were it not for a last ingredient: clay. It presents like microscopic platelets that stack on top of each other and ensure a solid cohesion once the water evaporates. The first constructions along the Nile and in the Fertile Crescent in Mesopotamia benefited from the immediate proximity of clay deposits found in riverbeds and in the subsoil.

The Shibam of the Animal World

Earthen constructions are not limited to human dwellings since many animals also use this method of construction. For example, small marine worms and honeycomb worms build mineral structures in the shape of honeycombs. They are made of sand and a glue they secrete, which is the key to their strong cohesion. All these cells constitute true reefs that spread over several hectares along a coast, as in the Bay of Mont Saint-Michel. Termites also build true earthen cathedrals several meters high with sophisticated towers and aeration shafts. These insects, true architectural experts in air conditioning, use a particularly fine clay that they mix with crushed mineral or vegetable waste. The resulting material possesses incomparable properties, so that abandoned, termite-mound clay is sometimes used in traditional ceramics or in the composition of coatings. Will tomorrow's (human) builders be inspired by these remarkable achievements to develop eco-friendly, biomimetic dwellings?

3 | *These natural, biogenic towers are made of digested wood and allow a natural control of temperature and humidity. This is of interest to entomologists ... and to architects.*

EXPERIMENT

You, too, can build with granular matter by making sandpies and testing their resistance to crumbling. To do this, make the pies the same size and vary their composition by using different building material. That way you will be able to compare more and less wet sands; or coarse sand and finely sieved sand. Try, too, to strengthen it using fibers, pine needles for example.

If you're into challenges, try to make a super-stack by filling a bucket with 1-centimeter-thick layers of sand separated by sheets of moistened paper (1), (2), and (3). On top of the structure you get, stack up several weights: a sandpie some 10 centimeters in diameter should be able to sustain tens of kilos.

4 | *This dry loam of sand was made by packing a mixture of sand and short fibers into a cup, layered as in rammed earth. It easily bears a weight of 10 kilograms.*

Materials

1.

2.

3.

Repeat x N layers → paper

LIQUID STONE: CONCRETE

Concrete has long been the cursed child of the construction world. Yet concrete's gray volume hides incredible scientific achievements. Rarely has any material made as much progress in so little time. Some contemporary architects have understood as much and appreciate the qualities of this liquid stone.

A Recipe Dating Back to the Roman Empire

No trip to Rome would be complete without a visit to the Pantheon. Its impressive 43-meter-diameter concrete vault makes it the largest dome of antiquity; and it is hard to imagine that the building has been used as it stands today ever since it was built under Emperor Hadrian twenty centuries ago. The mix of lime and clay had already been used earlier to

1 | *The immense concrete vault of the Pantheon in Rome has held up for twenty centuries. At the top of the hemispherical dome, an oculus nearly nine meters in diameter lets natural light into the monument.*

mix lime cement using sand. But it is the addition of pozzolan, a locally available volcanic sand, that would give this concrete its exceptional properties.

This amazing know-how, put to work in many buildings and described in Roman architect Vitruvius's *De architectura* in the first century CE, would unfortunately be lost with the fall of the Empire. Fifteen centuries later the process would be rediscovered and then improved in France by Louis Vicat, inventor in 1818 of cements that solidify under water. And concrete only acquired its modern composition with the Englishman Joseph Aspdin, a contemporary of Vicat, who filed the first patent for a hydraulic concrete (that cures with an admixture of water) and devised the Portland cement recipe, which has since become a worldwide standard.

"Dirty, heavy, and brutal," concrete has often had a bad reputation. And not for nothing; think of the ugly blockhouses left after World War II along the Atlantic coast of France, which required more than 10 million cubic meters of concrete; or the state of many immediately erected postwar constructions in Europe. The carbon footprint of concrete also argues against it—concrete is costly in energy: manufacturing cement contributes nearly 10 percent of the world's carbon dioxide emissions. Some of the billions of cubic meters of concrete used each year could certainly be replaced by natural, renewable materials [EARTHEN ARCHITECTURE]. But to readily reject concrete would be to forget too quickly that this material has embellished capitals such as Rome, which today owes its moniker of Eternal City partially to concrete. Moreover, thanks to recent innovations, concrete enables architecture to be quite daring. So, what if we rehabilitated concrete?

The Disadvantages of Concrete

Ordinary concrete has several disadvantages. Its porosity opens the way to stress caused by effects of its environment as well as the presence of residual water after it has set, all of which are causes of aging. On a mechanical level, one of the advantages of ordinary concrete is its compression strength, which is crucial in making vertical pillars. On a construction site, builders speak of 20-megapascal (MPa) concrete (a weight of 200 kilograms applied to a surface of 1 square centimeter), which is its colossal compression strength once cured. Conversely, conventional concrete does not withstand the tensile stresses to which it is subjected, especially horizontal structures such as lintels [AZAY-LE-RIDEAU OR ROOFS OF BEAUTY], which tend to bend at their center under their own weight and results in dangerously challenging their integrity. In the end, then, the cohesion of this granular amalgam is mediocre, a bit like sand [THE SECRET OF SANDCASTLES], which resists well only when compressed.

Concrete Revolutions

To increase concrete's tensile strength, in 1930, engineer Eugène Freyssinet (1879–1962) invented the *prestressed* concrete process. It consists in drowning taut cables in the mass of concrete while it's being poured. Once solid, the concrete is thus prestressed when the cables are released, thereby shifting the breaking point of the structure equally when stress is activated. Building sites inspired by Freyssinet's brilliant invention quickly spread across the planet: the first prestressed concrete bridge was built in Oelde, Germany, in 1938.

Nowadays the variety of concrete formulas and implementation compels us to speak of many *concretes*. Recent research has led to materials that are five times stronger than traditional concretes. These high-performance concretes make possible graceful and more elegant structures and decrease concrete's big carbon footprint. In Marseille near the Old Port, the MuCEM (the Museum of European and Mediterranean Civilizations) illustrates this (fig. 2). The building, designed by architect Rudy Ricciotti and his engineer son Romain, is alluring because of the lightness of its shapes, which are wrapped with a concrete and a porous fishnet-like mesh—creating a transparent interface with the city. The

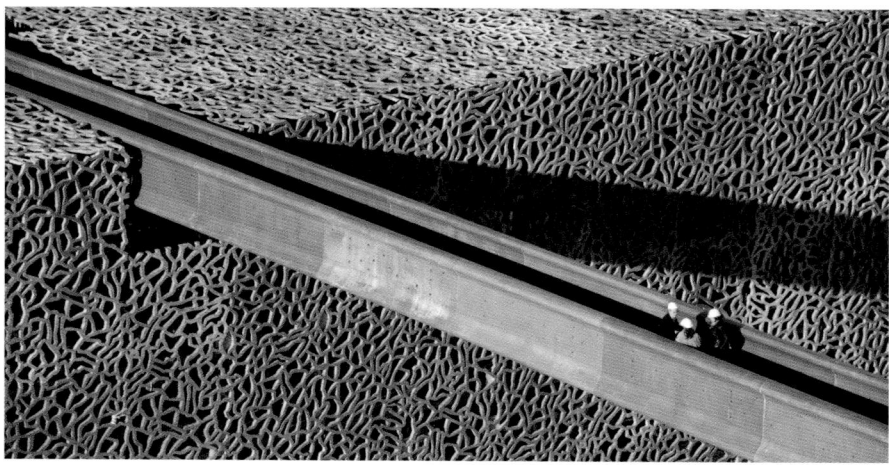

2 | *This extremely thin, 120-meter-long concrete footbridge connects the fishnetted MuCEM to the Old Port of Marseille.*

concrete is both its skeleton and its skin. More spectacular still, two foot-bridges connect the museum to Fort Saint Jean, which once upon a time guarded the entrance to the Old Port. Ricciotti and his son designed large concrete pieces prepared off site, which they assembled and tight-ened on site using metal cables kept in tension—much like building with architectural Legos!

Beyond aesthetics, this material has characteristics that the artistry of architects and engineers has magnified. The barely curved footbridge seems very thin, even a tad elastic, and it is difficult to believe that the concrete-veiled pedestrian way is only a few centimeters thick. What are the secrets of these new concretes?

Dust and Fibers to Consolidate

The composition of high-performance concrete (HPC) and ultra-high-performance concrete (UHPC) is one of the keys to this mystery. Ultra-fine particles smaller than 1 micron are added to ordinary concrete, mixed gravel, sand, and cement (fig. 3). These particles, which come from silica fumes, are interposed in the interstices of the pile of granular materials. And then what happens? They increase the compactness of the concrete. Moreover, thanks to the additional incorporation of fibers in the UHPC, the strength of the concrete can be further increased. In the end, these techniques allow for lighter and more durable structures that are also less sensitive to tensions.

Another advantage is that the amount of water in the concrete diminishes since the pores are now plugged by microscopic grains. Permeability, too, is greatly diminished and, with it, external stressors that are fatal in the long run. Exposed rebar sadly bears witness to this phenomenon. It doesn't seem unreasonable nowadays to guarantee that the numerous structures built with these new materials such as the Millau Viaduct or the Grande Arche de la Défense near Paris will exist for centuries. But the icing on the cake is that these concretes make it possible to create impermeable covers without recourse to additional treatments.

3   This simulation of the inside of a concrete sample illustrates the grain-size distribution, which ranges from gravel to sand and powders. Because it is polydispersed, it contributes to reducing space among the grains, which otherwise open doors to external stressors. Modern concretes are guaranteed for up to one hundred years.

EXPERIMENT

To test the importance of the granular organization of concrete with respect to its compactness, all you have to do is carefully fill half of a transparent jar with small granular material (sand or semolina) and then top it off with larger ones (chickpeas) (1). Now empty the contents carefully before thoroughly mixing (2) and (3), and refill the pot (4). Surprise! The total volume occupied by the mix has decreased.

You can also measure the compactness of the granular materials in the pots, that is to say the fraction of volume they take up. Once a pot is filled, weigh it; then add water up to the rim so that it fills all the spaces in between the grains; now weigh it again. The extra weight owing to the addition of water provides a measure of the void between grains. In relation to the initial volume of the pot, this volume indicates the porosity of the granular stack. The porosity is greater when there are two superimposed piles of grains of the same size.

Materials

1.

Fully full!

2.

3.

4.

5.

Fully compact.

Not so compact.

THE SAGA OF FUSING GRANULAR MATTER

Who would suspect that sublime Chinese porcelain actually hides a subtle phenomenon whose secrets physicists have not yet uncovered? This is the art of binding powder grains together by baking them.

The ceramics (Greek keramos, *"clay") obtained this way are inorganic materials that—from porcelain to the fragile kitchen knife that can't be worn out—are applied to many different areas.*

These Chinese cups are decorative, aren't they? For anyone who takes a close look at them, their porcelain is absolutely graceful. An amateur will appreciate the brilliance, the whiteness, even the hardness of a beautiful porcelain that can be tested by making it ring with a small spoon. Coming from a long line of high-tech products dating back to the sandstone pots

| 1 | *A blue and white Chinese porcelain cup from the seventeenth century. Porcelain is a ceramic made of very small, fused kaolinite grains.* |

that appeared in China over twenty centuries ago, this supremely luxurious china would be exported all over the western world. The recipe for making it would be leaked by a Jesuit who had lived in China, enabling porcelain to be produced in Europe starting in the eighteenth century.

Porcelain is highly sought-after because it can be shaped easily and then processed to give it its exceptional hardness, so that the sharpest knives leave no mark on it. How is that possible? The answer lies in the structure of the material and in its manufacturing method.

Porcelain Recipe

Porcelain is the first example of hard material obtained by fusing grains. It is made from *kaolin*, a refractory and friable white clay. The first deposits in France were discovered near Limoges, which became France's porcelain capital. Kaolin melts at 2000 degrees Celsius but, by adding an adjuvant (feldspar), the temperature at which a precursor of porcelain is formed drops to 1300 degrees Celsius.

At a microscopic scale, each grain of this mineral resembles a tiny mille-feuille with a hundred layers consisting of many sheets just nanometers thick (fig. 2). These layers, which can slide against each other before being fired, allow the shaping of the dough. We speak of plasticity to account for this deformation. A first firing of the cup or plate fuses the grains by eliminating the water that was between the layers. After this step, the material remains porous. It is then covered with a solution of a metal oxide that fills the pores. Its surface will take on its beautiful shine once the object is put back in the kiln. During the second firing, the clay continues to transform and leads to the final constituents of porcelain made from alumina and vitreous silica.

A Cement without Glue

Cousins of ceramics that are obtained by firing clay are *sintered* materials resulting from the hot fusion of a powder without resorting to an external "glue" such as a cement [LIQUID STONE: CONCRETE]. Before firing, the powdery solid is put into a mold that will endow it with its final shape. Heated at a temperature lower than that of bulk melting, the atoms close to the surface migrate to the areas among the grains and connect them; it is this material among the grains which carries out the fusing.

> 2
>
> The kaolinite platelets—the mineral species found in kaolin—seen under a scanning electron microscope: they form a fascinating, disordered gathering. These platelets are one-hundredth to one-tenth of a millimeter long. This layered structure explains the plasticity of the clay before firing: clay leaflets can slide easily over each other.

Sintered materials used today in industry, such as zirconia or alumina, combine high mechanical rigidity with lightness; alas, like other ceramics, they remain brittle. They were developed in two opposite directions: making very porous materials or very compact ones.

Sintered with Holes, like Swiss Cheese

Fusing granular matter makes it possible to create hollowness out of fullness. Sintering is used to design highly porous materials for the aeronautical industry in order to lighten airplanes. Nowadays, surgery and dentistry also make use of it widely: porosity allows compatibility with cancellous bone and promotes the growth of tissues at their interface, to the point that implants end up completely embedded in the bone (fig. 3).

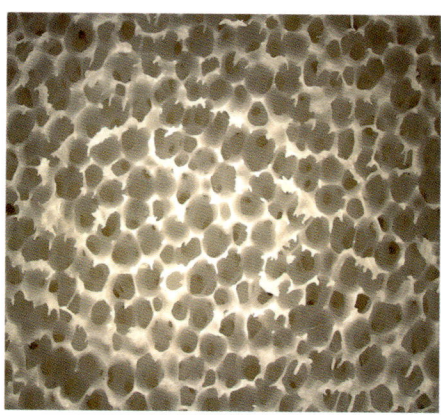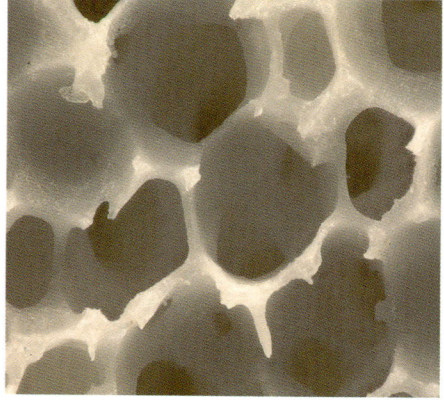

3 Enlargement and detail of the surface of a piece of sintered alumina (with 60 percent porosity) used in prostheses. This structure reproduces that of the cancellous bone (the size of these pores is about 1 millimeter in diameter).

Durable Knives

Sintering also leads to very compact ceramic materials, obtained by applying an intensive treatment that eliminates the pores between the grains of material (fig. 4). They are mainly used in some mechanical parts and for making knives. The latter are made from a powder of metal-refractory oxides, previously pressurized and then heated to 1500 degrees Celsius. Such cutlery does not rust and is harder than steel; sharp as glass, they are also brittle like glass.

Through the diversity of materials obtained by fusing granular materials, one can thus obtain very varied geometries. The sintering phenomenon is used today for 3D printers, which are already revolutionizing traditional factory and machining technologies.

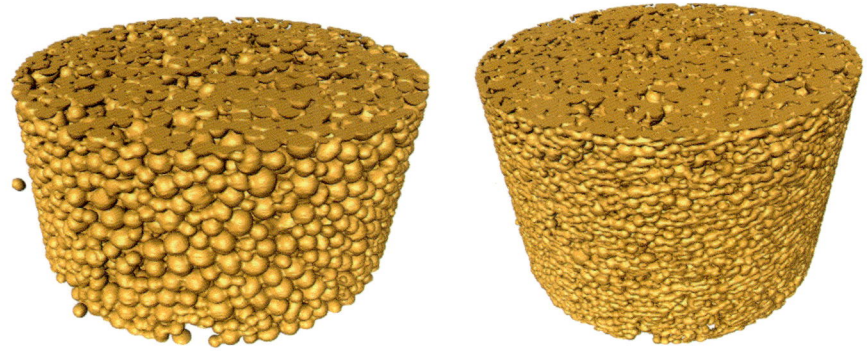

4 | *These volume images of a sintered copper sample were obtained by synchrotron radiation. On the left, the sintering is weak, as shown by the bridges (or necks) between grains, and the decrease in porosity is limited. On the right, the sintering is strong, the grains are very deformed, and the porosity largely reduced.*

EXPERIMENT

What makes a snowball hold together? What if you spent time over your winter break extending the debate between Lord Kelvin and Michael Faraday? These two great nineteenth-century English scientists looked into the problem of sintering the ice crystals that make up snowflakes. Would molecular films of water be spontaneously present on the surface of snowflakes? They might thus lead to capillary bridges that, once solidi-fied, would ensure the cohesion of snowflakes. Would pressure applied locally melt the crystals at points of contact? Or does this cohesion simply stem from the entanglement of crystals?

None of the solutions they proposed is entirely satisfactory: snowballs can be formed at minus 30 degrees Celsius, in the absence of any liquid water. In addition, the pressure applied between the grains is not enough to melt the ice. There is ongoing research trying to solve this mystery.

To make up your own mind, get a hefty amount of fresh snow and start by packing it into a snowball. Does liquid water get expelled, like from a wet sponge? Not really. Test the strength of the compacted ball by drop-ping it to the ground from a given height. Does it break up at impact? We can intuit that a snowball will be all the stronger if it has been heavily compacted. How much can you compress your snowball? What are the physical ingredients that explain cohesion?

STATES OF GLASS

So common and yet still fascinating, the beauty of glass has amazed human beings since prehistoric times (glass tools dating from the Neolithic era have been found). No doubt magical powers were attributed to this natural glass, as evidenced by the beetle that adorned the breastplate of young Tutankhamun. It is less known that glass remains a mystery for science. Both brittle and elastic or, looking closely, solid or liquid, it combines surprising and still imperfectly understood properties.

What is more commonplace than glass in our everyday lives? It comes in such a variety of shapes, colors, and uses that we hardly take the time to admire its strangeness (fig. 1). To arouse our curiosity, let's visit Jeremy Maxwell Wintrebert's studio in an arcade on Avenue Daumesnil in Paris (fig. 2). This artisan creates huge works made from glass. At the end of a long steel rod, he has just taken a pasty mass of glass out of the furnace,

> 1 | *The charming linked islands of Murano (near Venice) have been the home of glass artists since the fourteenth century. Although initially transparent, the molten glass used in Murano vases is tinted by adding metal oxide: cobalt for blue, copper for red, and iron for green.*

at a temperature exceeding 1000 degrees Celsius. With a quick gesture, he turns it continuously to prevent it from falling.

Working rapidly, the glassworker blows into the blowpipe to shape the glass. At the other end of the pipe, air starts to create a bubble inside the hot and viscous mass, which grows into a beautiful, incandescent sphere ready to be shaped. Each creation requires great precision: working a few seconds too long or heating a few degrees too much can lead to a resounding failure. What is the physical principle that creates this magic?

Shaping Glass: From Obsidian Blades to Marbles

Vitreous materials have been identified and exploited since the Neolithic era. There even developed an important trade in blades carved out of obsidian rock, which is of volcanic origin [PREHISTORIC GEMS]. More surprising is the origin of Libyan glass from which Tutankhamun's beetle ornament was cut. This glass seems to have been produced when a meteorite exploded in the atmosphere above the Libyan desert nearly thirty million years ago. The extreme heat it generated would have caused the sand to melt into glass.

Fortunately, we no longer need to wait for geological disasters to obtain glass. The major constituent of glass is silica, which is essentially sand. It liquefies at 1700 degrees Celsius, but this melting temperature can be lowered around 1000 degrees Celsius by adding so-called fluxing agents (most often potassium oxides or calcium oxides). In a furnace, glass flows like honey. When it is taken out, its temperature gradually decreases, which leads to its increased viscosity: it becomes pasty at around 900 degrees Celsius.

It is during this slow cooling down that glassblowers shape their work, taking advantage of its viscous, liquid state. They use tongs, plates, and scissors. But they must not take too long because at 500 degrees Celsius there's no way to work the glass any longer: it will indeed have gone from a hot, ductile, and malleable state, in which it deforms easily and irreversibly, to its cold, brittle, and fragile state. A birthday candle immersed in hot water behaves the same way: when hot, it continuously deforms in a ductile manner and becomes breakable when cooled off.

When you've played with marbles, have you ever wondered how they were made (fig. 3)? The same trans-formation takes place: a glass thread heated to 1200 degrees Celsius is first cut into small cylinders, then mixed with colored globs, and then rotated in a rotating container. By dint of rolling around, the glass glob gradually rounds out and finally adopts a perfectly spherical shape while cooling.

A Liquid That Doesn't Know It's Liquid?

But why didn't the glass solidify at a specific temperature, just as water goes from a liquid state to a solid state at 0 degrees Celsius? And, besides, at this temperature two phases exist: liquid water and ice. What explains the continuously changing mechanical properties of glass as it cools down? Admittedly,

> 2   *Glassblower Jeremy Maxwell Wintrebert has only a few dozen seconds to use his scissors and shape this imposing mass of glowing glass at the end of his rod. An assistant constantly turns the rod to prevent the viscous glob from collapsing under its own weight.*

| 3 | These glass marbles were made by rolling cylinders of molten glass paste during its transformation from a ductile state to a fragile state. |

its viscosity increases sharply during the cooling phase, but without any discontinuity. There's no noticeable reorganization at the microscopic scale, unlike water, which undergoes a sudden transition from a disordered liquid state to an ordered crystalline state. That's a major difference: there is no way to blow ice the same way one can blow glass!

In fact, glass is a vitreous material, and its mineral constituents are in total disorder. These molecules, which are extremely agitated at high temperature, gradually come to rest when cooling down, keeping their disordered arrangement. To understand the difference when water changes states, one can compare an ice crystal to a pile of oranges neatly stacked in grocery store; to imagine the liquid state, one must imagine the same batch of oranges knocked around and all over the place in a grocery truck on a bumpy road. So, therefore, is glass a liquid that doesn't know it's a liquid? Not quite. Certainly, at the molecular level, its structure matches that of the liquid's instantaneous disorder. But unlike a liquid constantly shaken by thermal agitation, this disorder solidifies at low temperature. Thus, the glass, at an ambient temperature, behaves like a solid.

Water, a Great Enemy

And yet this solid is brittle: small defects such as microscopic scratches on the surface of a piece of glass can easily propagate and give rise to large-scale cracks. Only glasscutters take pleasure in the fragility of glass when cutting glass panes. It is the scratch of a tungsten carbide cutter that makes it easier to cut glass cleanly.

The possible presence of moisture is another concern in industry: when water molecules come into contact with the tip of a crack, they promote breakage. They attack the silica by hydrolysis, which allows the crack to progress as if a microscopic zipper were opening. To reduce harmful scratches, one solution is to treat the surface. One might, for example, direct heat to a specific portion of the glass to melt it and even out its surface.

Another way of proceeding is to protect the surface of the glass, a method used with fiber optics, those long glass wires that connect us to the internet and to telephone communications across the continents. Highly mechanically stressed during their handling, these optical fibers are covered with a polymer film that protects them against mechanical and environmental damage. The endlessly long glass cylinder, sheltered in its sheath, thus can be reversibly bent to astonishing degrees without breaking, which facilitates their handling and storage.

Metallic Glass

Not all glass is made of silica. Have you heard of metallic glass? An amorphous glassy metal can be obtained by rapid cooling from the molten state: a metal's atoms are agitated and disordered in their liquid phase, and they are then solidified while remaining disordered. At low temperatures, a solid metal is usually composed of small, entangled crystals; these structural defects are in large part responsible for the mechanical properties of the final material. In amorphous metals, it is the whole of the material that is disordered. As a result, such *metallic glasses* can be elastically deformed to much higher values than ordinary metals. Paradoxically, it is the atomic-scale disorder that reinforces the material!

EXPERIMENT

Simple melted sugar is enough to simulate how silica glass is made. In a kitchen container, mix 300 grams of table sugar (sucrose), 50 grams of glucose, and 75 grams of water, which you will melt by heating the mixture to a maximum temperature of 150 degrees Celsius (be careful not to burn yourself, and do not heat it too long or it will turn into caramel) (1). You can use isomalt instead of a sugar-glucose mixture—a commercially available sweetener that does not turn brown when cooked.

During the cooling phase, pull out a strand of the mixture using a fork (2). Drop it in cold water (3): you will make the equivalent of fiberglass. You can also spread a part of this preparation on a plate; it will soon become a soft paste, which you can then shape like molten glass. When cold, this paste is more or less a vitreous layer. Note that without adding the glucose, the same experiment would yield crystalline sugar, which is impossible to mold. Like glassblowers, some confectioners have become masters of shaping sweet analogues. Can you match their artistry?

4 | *Blown-sugar figurines are popular in Asia. The body of each figurine was made by blowing through its hollow tail used as a straw. The final shape and the limbs have been molded by hand (the temperature of melted sugar being much lower than that of glass).*

Materials

ISOMALT

1.

2.

3.

V

MATTER IN
MOTION

*Like this fiddlehead just waiting to uncurl, plants are
capable of moving in ways whose secrets we are just
beginning to grasp. On closer inspection, their placidity
hides mechanisms of incredible violence and speed,
such as those that ensure the ejection of seeds. Plants
are not alone in exploiting elastic structures in order
to move: a violin string and a pole-vaulter's pole are
examples of items that you can load and discharge.*

BEAMS BEND BUT DO NOT BREAK

What is the relationship between a flower in a vase and the terrible collapse of the World Trade Center twin towers in 2001? The answer: an impressive mechanical instability—the architect's scourge.

Erect and resplendent, Sunday flowers decorate your living room for a few days, and then one morning you see one bending under its own weight. But why is it suddenly bent over?

> **1** *What happened to this wilting gazania bending under its own weight? It had earlier looked so proud.*

In the Plant World ...

Like any structure on Earth, a flower must fight against gravity to grow upward. Although its stem is scrupulously vertical, its balance is only possible at the cost of compression leading to an imperceptible reduction of its height. But if the stem tilts a bit to the side, then its weight acts as a lever that tends to make it lean over even more, which in turn increases the leverage, and so on. This vicious cycle is what physicists describe as buckling.

How does a flower initially manage to resist buckling? Resistance to bending counters the stem's instability. In a freshly cut stem, cells are still considerably swollen with water; their internal pressure then is several times greater than atmospheric pressure, so the stem remains taut like a well-inflated bicycle inner tube. But these cells die off over days; they can no longer maintain the water they have been absorbing in the vase; and they gradually lose their internal pressure. Having become much more flexible, the stem can no longer withstand the destabilizing effect of its own weight and folds over.

... and What about in Human Constructions?

Buckling instability is universal: it concerns almost all slender structures and appears when they are sufficiently compressed. A compressed beam thus remains straight as long as the force exerted on it is weak, but it bends as soon as the load exceeds a threshold value.

This buckling instability threatens stiff structures. This is the case of rails that are compressed in an earthquake (fig. 2). They would no doubt have withstood the impact had they not been compressed but stretched

2 | *During an earthquake in New Zealand in September 2010, railways were so compressed that they finally bent and created a kind of wave shape.*

instead with the same strength. The phenomenon is both treacherous—below the threshold, few signs foreshadow the imminence of the deformation—and reversible: in the end a bent beam regains its straight form when the compression applied is reduced.

Beyond the threshold, however, bending increases quickly and can lead to irreversible deformations and even ruptures. Fortunately, the operating regime of our buildings is very much below the threshold: engineers are not taking any risks with bridges that might collapse under heavy traffic! Nevertheless, the sudden collapse of the World Trade

Center's twin towers in New York on September 11, 2001 suggests that the heat generated by the burning airplanes considerably reduced the strength of the structure's vertical steel beams. The buckling instability of internal beams led to the violent collapse of the upper level, eventually causing the collapse of the whole building.

Ruining One's Car rather than One's Life

Contrary to this buckling, a priori a progressive phenomenon, there is a related instability that is intrinsically abrupt and irreversible. In nature, some plants resort to such sudden instabilities to produce a rapid movement at the end of the summer to eject seeds [FLYING SEEDS].

When mechanical engineers compress a hollow cylinder, they demonstrate spectacularly what they call structural ruin. Carefully place your foot on an empty, vertical aluminum can: it will probably support the pressure by deforming very little. But ask an assistant to flick it on its side, and it will suddenly crunch.

In the automotive field, this kind of dramatic instability is exploited to improve passenger safety. In a violent crash, a car's peripheral structures are designed to be crushed like an accordion. This irreparably damages the body of the car while absorbing a colossal amount of energy—which preserves lives.

Multiple Bucklings

When you think about it, our daily life is teeming with examples of this type of stark instability, starting with … foam mattresses, where this phenomenon paradoxically helps us sleep better. They are in fact made up of numerous fine-walled cavities, each of which behaves like a can that abruptly gets crushed. Their walls, however, are much more elastic than the aluminum of beverage cans, so the crushing is reversible—and that's a happy thing.

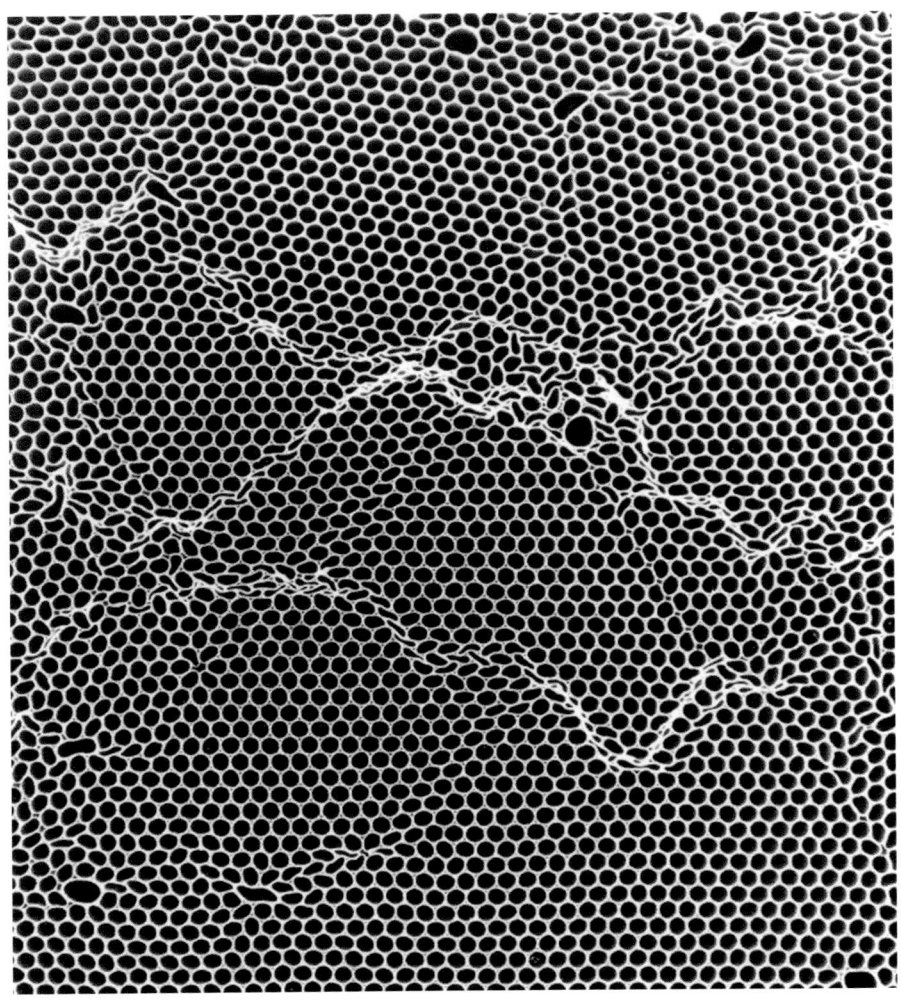

3 Plastic straws, whose ends we see here, were stacked regularly and then subjected to vertical pressure. Curiously, the crush is not uniform: some lines of straws collapsed (in white on this image), while others are hardly deformed.

But when we lie down on the mattress, only one part of the cells flattens, which has the effect of maintaining a constant pressure under our entire body. The mattress conforms to our shape without increasing the pressure where it is most deformed: our weight is thus distributed, in the end, affording greater comfort than old beds with box springs did.

An elegant experiment illustrates this mode of instability, using a model mattress made of stacked, parallel straws (fig. 3). A light load causes a slight compression of the straws. But, beyond a threshold of pressure, a line of straws suddenly collapses. By barely increasing the load, other lines soon collapse, too. Just like your trusty foam mattress, this lab model crushes at virtually constant pressure.

The unavoidable and sudden nature of the instability of crushing has not been lost on contemporary artists, who have made use of it in their compositions. The works of Edmond Vernassa from Nice, France, thus present an art under constraint, like this metal drum that dramatically collapsed when the air inside it was pumped out.

4 | *This work by contemporary artist Edmond Vernassa is on exhibit at the University of Nice. It was created by gradually by pumping the air out of the drum before it suddenly imploded. This caused a noise so deafening that Vernassa swore not to repeat his experiment, attesting to the violence of certain instabilities.*

EXPERIMENT

To illustrate the phenomena of sudden instabilities, horizontally unroll a metal tape measure with the tape measure's concave side facing upward. At first, it stays horizontal, bending only slightly. Soon, beyond a certain length, it will bend suddenly with an audible, cracking sound. Then retract the tape a bit: you will notice that it remains folded over. Such sudden instability is not reversible. In fact, at this length the tape measure can be observed in two states: almost straight (1) (which can happen again only after slightly rewinding the tape measure) and slouched (2).

The tape measure's stiffness is caused from the section's curved shape; its depth gives it a significant apparent thickness, but also suggests the cylindrical shape of our earlier soda can and its dramatic behavior when crushed. In architecture, many structures such as I-beams and hollow cylinders are reinforced by shape effects following this principle. The price to pay sometimes is the risk of a sudden deformation in the case of an overload, as we just saw with the tape measure.

Materials

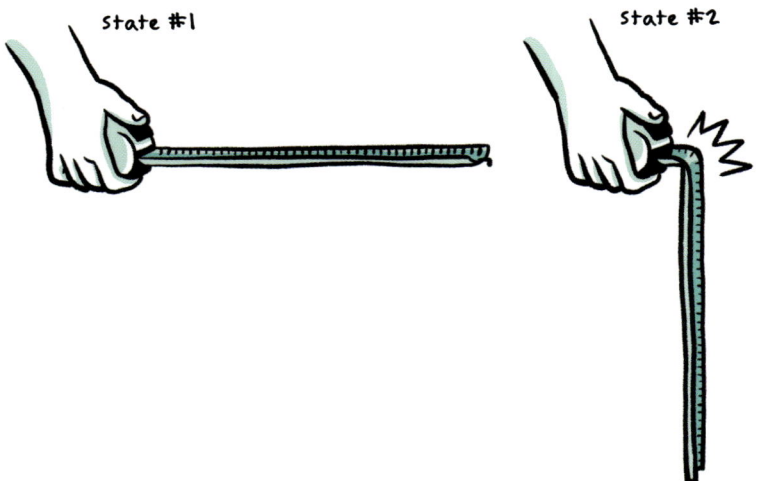

State #1

State #2

215

VAULTING INTO THE AIR

Pole vaulting illustrates how to combine "gravity and grace," to borrow from the title of a work by philosopher Simone Weil. Not to mention the thrill as the pole bends so dramatically just as it is about to propel the pole-vaulter into the air. The state-of-the-art materials that now make up vaulting poles partly explain ever-improving performances in recent decades.

Athletes never fascinate us as much as when they compete with minimal equipment. Pole vaulting particularly embodies such naked simplicity that highlights the physical beauty of athletic feats. At the mention of Sergei Bubka or Renaud Lavillenie one can immediately visualize their bodies, which spring off the ground, rise feet first, and cross over the bar barely brushing by it.

1	*The energy built up during previous world record holder Renaud Lavillenie's run-up is converted into the bending energy of the pole and gathered to propel the athlete upward.*

The present world champion, the Swede Arnaud Duplantis, vaulted 6.18 meters in 2020, gaining a few more centimeters on Lavillenie's record, by running even faster before his jump. This record will doubtless keep being broken! But how does an athlete manage to reach the sky? What is the physics involved?

The Physics of Jumping

Pole vaulting has a feature in common with the high jump: the approach run. Performing a standing high jump (which used to be an Olympic sport), one can barely reach 1.5 meters. During the approach run, the jumper progressively increases his speed and accumulates kinetic energy. During the jump, that energy is transformed into another form, namely

2   *Étienne-Jules Marey's 1890 chrono-photography is a sequence of shots taken a fraction of a second apart and superimposed on the same negative. We are seeing the end of the run (from right to left) and the vault of an athlete using a stiff, 3-meter pole. This traditional technique, without turn, leads to a relatively modest performance.*

potential energy, which is calculated by multiplying the athlete's weight by his height. The height that matters here is that of the athlete's center of gravity, located at the average position of the mass of his body. Vaulters and jumpers understand the physics well: they curl their bodies around the bar precisely in order to keep their center of gravity low and thus gain precious centimeters.

Storing Kinetic Energy

Whether the high jump or pole vault, a crucial moment is when the takeoff foot loses its ground support: the whole challenge is then to convert the momentum gained during the horizontal run into vertical displacement. During a high jumper's approach run, the athlete reaches a speed of about 10 meters per second, close to that of a 100-meter sprinter. If he were to hypothetically use all of the stored, kinetic energy to take off, a quick calculation shows that he would clear 5 meters at best: that's the height a ball would reach when tossed upward with the same initial speed. With high jump records of around 2.40 meters, athletes are still far from this theoretical limit, which means that the energy conversion remains very incomplete. While a pole enables better energy conversion, it would be pointless to hope to jump much higher using a much longer pole. Pole vaulters, however, do exceed this 5-meter barrier. Are they defying the laws of physics? In fact, once the pole is planted, these athletes use their arms. The downward thrust exerted on the pole during the vault provides valuable extra energy, which in turn gives the vaulter a solid, extra meter.

The Secret of the Poles

Pole vaulters' results have almost doubled since the American William Hoyt's 3.30-meter jump at the first modern Olympics in Athens in 1896. How is such progress achieved? Let's take a closer look at the pole.

Olympic rules do not regulate constraints on the pole's material or length, given that the pole's surface must simply remain smooth. It is this freedom that has made it possible to explore very varied solutions, motivated by the race for new records.

Early poles—made of wood, bamboo, or metal—barely bent, as shown in Étienne-Jules Marey's famous chrono-photograph (fig. 2). Contemporary poles made of composite materials appeared at the 1964 Olympic Games. Such material combines reinforcements made of glass or carbon fiber, which are responsible for flexibility and resistance, with a polymeric resin, which ensures the cohesion of the whole while keeping the pole's weight down (3 to 5 kilograms). This flexibility has revolutionized the body techniques of pole vaulting.

The pole has two functions. First of all, as we have seen, it redirects the horizontal speed of the approach run to a vertical speed. In addition to this, when the vaulter plants the free end of the pole in the box, he increases the bend of the pole in the final phase of his run: the elastic energy stored by the pole is then gradually restored to the jumper and allows him to rise. The numerical modeling of a jump's different phases emphasizes the essential mechanical function of the pole's elasticity. Questions arise for researchers in mechanics: does a pole that is too long run the risk of buckling and failing [BEAMS BEND BUT DO NOT BREAK]? Is it possible to use the pole's vibrations to gain more height? On the one hand, if the pole is too flexible, the energy release is not carried through quickly enough and the transfer is inefficient. On the other hand, if the pole is too stiff, the impact as it makes contact with the ground results in a loss of energy; besides, the pole vaulter is thrown back. In the later phase of the vault, the pole also serves the vaulter, once his body is upside down, to rise higher using the strength of his arms. Again, a pole that is too flexible would be unable to produce this additional impulse. Future champions will also be fine mechanical engineers!

The pole acts like a catapult: the energy stored during the approach run is used to *gradually* bend the pole when it is pressed against the box. During the vault itself, the bend disappears and the mechanical energy

3 | *The joints of grasshoppers' hind legs have a mechanism similar to that of a catapult or an archery bow. It aims to suddenly release the energy grasshoppers store when they fold their legs.*

accumulated in the flex is transformed into potential energy as the vaulter is propelled upward: this release must be accomplished quickly to give the vaulter maximum impulse.

This catapult effect is a very general phenomenon, which grasshoppers, for example, use too. It is by folding their long legs backward that these animals gradually store elastic energy. A quick release activates a dramatic unleashing mechanism that relaxes its legs at a speed its muscles can't achieve. They jump as high as one hundred times their size. An analogous mechanism is at work in rubber poppers [FLYING SEEDS]. You may have also used the same mechanism without knowing it by snapping your fingers; the noise produced comes from releasing your thumb very quickly.

EXPERIMENT

If you press too hard on a beam, it bends and eventually breaks. Richard Feynman (1918–1988), Nobel laureate for his work on quantum electro-dynamics, was known for his scientific eclecticism. Thus, he asked the following question: Into how many pieces does a piece of raw spaghetti break?

Try it in your kitchen! Take one piece of spaghetti, hold it with one hand at each end, and force it to bend gradually by rotating your hands. It will eventually break up. How many pieces do you get?

Contrary to what you might expect, it doesn't usually break into two pieces, but rather into three or more. Why?

French physicists have answered this question: using a fast camera (several thousand images per second) they filmed the breaking apart of a piece of spaghetti flexed at its ends. This study, which is much more serious than it seems, earned its authors an Ig Nobel Prize, a prize awarded to unexpected, surprising, or paradoxical research. After a first break, the piece of spaghetti is traversed by a flexion wave that increases its curva-ture locally: it is this wave that is likely to cause the piece of spaghetti to break once again.

The illustration's sequence of photos shows how the wave propagates along the beam. It ends up being reflected at the beam's end, which leads to increasing the bend all the more as the wave that leaves and the part of the wave that has not yet arrived interact. This pronounced bend pro-duces a break. Finally, the spaghetti's multiple fractures do not take place simultaneously, but one after another. This mechanism is not only true of spaghetti: when a pole vaulter breaks his pole it often breaks into three pieces, as can be seen on many internet videos.

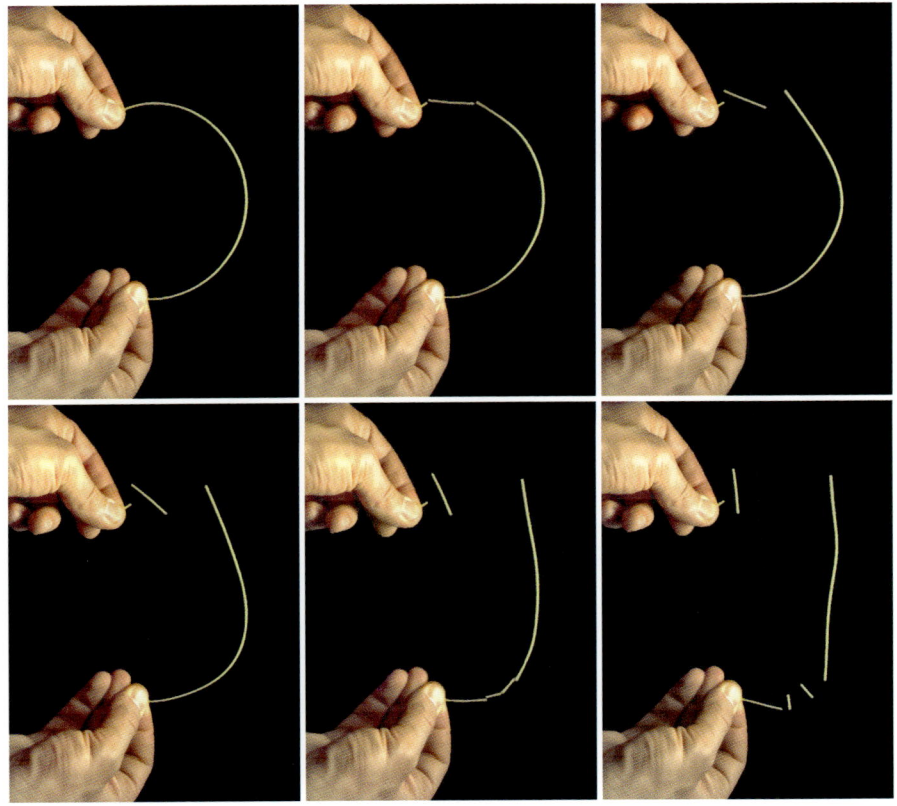

4   *A sequence of images illustrating the propagation of the bending wave at the instant when a piece of spaghetti breaks under too much bending. This wave is reflected at the end of the piece of spaghetti and interacts with the part of the wave that has not yet arrived, which amplifies the local curvature and induces new breaks (one-thousandth of a second between the images).*

THE CHOREOGRAPHY OF PINE CONES

Did you know that to release their seeds, pine cones open and close thanks to an ingenious mechanism controlled by ambient humidity? Other plants are also set in motion by simple external stimuli. These elegant solutions are now used for bio-inspired applications.

His face appeared to me as long and large as is at Rome the pine cone of Saint Peter's.
> —Dante, *The Divine Comedy, Inferno* XXXI, vv. 58–59, trans. Allen Mandelbaum.

The Cortile della Pigna, an inner courtyard that visitors doubtless encounter as they tour the Vatican, takes its name from the colossal statue in the shape of a pine cone erected in the niche of the Belvedere Palace. The origins of this work are not known with any certainty; some think it adorned the dome of the Pantheon. The numerous holes located among

| 1 | *The pigna, a bronze statue in the shape of a pine cone, is housed in the niche of the Belvedere Palace in the Vatican.* |

the scales of this conifer suggest that this bronze work served as a fountain in Roman times, perhaps near the Baths of Agrippa. That the Romans should glorify an ordinary pine cone is puzzling if you don't know that in antiquity pines symbolized immortality because of their evergreen foliage. An essential raw material, this tree was used for shipbuilding at the time, while its resin made it possible to seal hulls and preserve wines.

Although pines give us a feeling of immutability, such an impression is misleading. Far from being inert, their fruits perform a peaceful dance

2 | *Saturated with water (left), a pine cone remains closed. When drying, its scales open, which releases its seeds and allows for them to disperse.*

at a tempo dictated by the weather. Indeed, you may have noticed, walking through the forest, that mature pine cone scales close up in wet weather. When the sun returns, they dry and reopen (fig. 2). Their deployment makes possible the release and dispersion of seeds trapped between their scales. But how can a piece of wood be the site of such movement?

Curvature Leads the Dance

Simply cut a scale in half via its plane of symmetry (fig. 3) to understand the basis of these hygroscopic moves: a *bilayer* structure appears. One of the sides of the scale is essentially made of a type of wood that swells with moisture. Here's how: water molecules bind to cellulose fibrillae in the walls of plant cells. The distance between the fibrillae then increases slightly, which on a macroscopic scale translates into the wood swelling around 10 percent. The water's attachment to the cellulose is reversible: in a dry environment, the water subsides, and the material contracts and returns to its original shape.

The other layer forming the scale is passive, that is to say, it does not swell in the presence of moisture. It is the differential expansion of the two materials that alters the curvature of the scale in response to fluctuations in humidity (fig. 4). Pine cones are natural hygrometers, which has the advantage of not requiring of the plant any active control or energy supply once the pine cone has matured.

Ferns and Jack Pines

In addition to the pine cone, the plant world is rich in structures whose shapes change with changes in humidity. The seed of *Erodium cicutarium*, a herbaceous plant with purple flowers, winds itself into the form of a helix thanks to hygroscopic deformations, which allows it to sink in the ground like a tendril. As for the spores of some horsetails, their branches open and close with moisture, which allows them to move on the ground, or even to jump in order to blow in the wind.

Another example: when subjected to drought, pollen grains tend to fold upon themselves, which reduces their desiccation. Finally, jack pines have an astonishing peculiarity: their cones, bursting with resin, open only when a forest fire occurs. Then they inseminate the burned soil first, giving them a definite advantage over species that will come later. This phenomenon also affects sequoias, and a controlled fire strategy has been put in place to favor them. Whether these bilayered strips are hygroscopic or thermometric is a question that is not completely settled today.

From Sportswear to Thermometers

Inspired by pine cones, it is possible to take advantage of a material's differential deformations to make sensors or objects driven by variations in humidity. A slip of paper glued to a thin, plastic blade thus constitutes a cheap hygrometer. Some sportswear manufacturers produce adaptive fabrics containing bilayered scales: they open to let transpiring skin breathe. Why not imagine a shelter unfolding itself when it rains, a nonskid floor covering for wet weather, or a system that captures the energy of humidity fluctuations in the environment?

The same sort of mechanism is at the core of temperature probes consisting of two different, superposed metal plates. Depending on their nature, these metals expand differently as temperature increases, their differential expansion causing a change in curvature. This mechanism is the active part of many devices (thermometers, circuit breakers, thermo-

3 | *Cutting a pine cone scale across its plane of symmetry reveals a bilayered structure at its base (at the bottom of the image). The dark outer wood (lower part) swells noticeably with ambient humidity, while the lighter wood (upper part) is much less sensitive.*

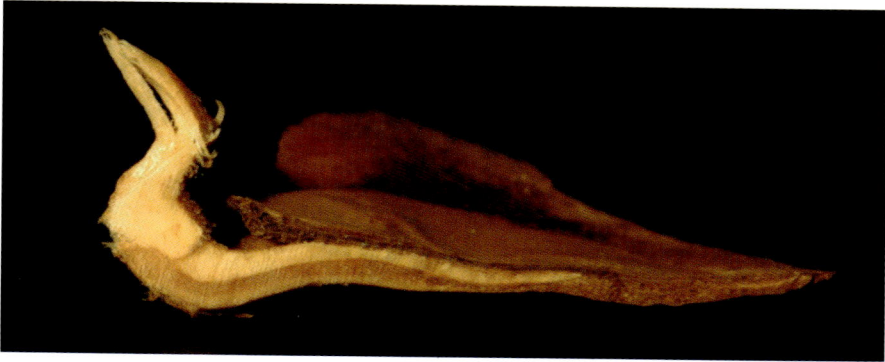

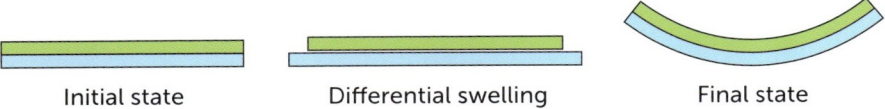

Initial state   Differential swelling   Final state

> 4   *Principle of bilayer mechanics. Two blades are laid one on top of the other. One (in blue) swells when it comes in contact with water, the other does not. Once the blades are glued, the spontaneous curvature makes it possible to accommodate the lengthening of the first while respecting the weaker dilation of the second.*

stats, flashing lights, etc.), capable of either switching off under the effect of heat, or alternating between *on* and *off* positions. In the latter case, the bimetal bends because of the temperature, breaks the circuit, cools, and returns to rest, and then the cycle starts all over again.

> 5   *Detail of the hygroskin pavilion envelope (permanent collection of the FRAC Center Orléans). Formed with bilayered strips, the openings react to moisture fluctuations, thus modulating the aperture, the texture of the surface, and the transmission of light.*

EXPERIMENT

"The Japanese amuse themselves by filling a porcelain bowl with water and steeping in it little crumbs of paper which until then are without character or form, but, the moment they become wet, stretch themselves and bend, take on colour and distinctive shape, become flowers or houses or people, permanent and recognizable." As Marcel Proust invites us to do in *Swann's Way* (Scott-Moncrieff's translation), let's play with paper. Cut out a sheet of tracing paper, about 5 centimeters square. Lay it gently flat on the surface of a small tub of water. You will see that in a few seconds the paper curls up into a millimeters-tight roll. In fact, it temporarily adopts a bilayered structure: in contact with water, the underside of the paper tends to swell, while the upper surface remains dry, at least initially. When the paper is fully and evenly wet, it reopens slowly.

To extend this experiment, cut several strips from your sheet of tracing paper, one along the long side of the sheet, the other along the short side, and finally a third diagonally (1). Curling occurs parallel to the paper fibers, which have been aligned during its manufacture.

The game evoked by Marcel Proust is different (it is called *suichuka*). Cut a square from a sheet of newspaper and fold over the four corners so that they meet at the center (2). Place the folded-up square on a water surface. Unlike with tracing paper, you will not see any curling, but instead you'll see the folds opening spontaneously. Now try other configurations by coupling cutting and folding, in the manner of artist Étienne Cliquet and his *Flottilles*, which you can see on the internet.

How does this unfurling happen? In the case of newsprint, soaking happens very fast so that the bilayer effect does not have time to occur. The paper swells throughout its thickness, and the inside of the fold, initially compressed, relaxes.

Materials

1.

2.

FLYING SEEDS

We've all blown on dandelions to watch their achenes fly and their elegant way of drifting like small parachutes. This may seem like frivolous play, and yet, for the plant, scattering its seeds is a serious matter. Some use the wind, water, or birds flying by; others have developed ingenious, fast ejection mechanisms. The diversity of processes with which evolution has endowed the plant kingdom is spectacular.

A carnivorous plant can't be vegetarian, at least I think not.
—Jean-Marie Gourio, *Brèves de comptoir.*

The plant world is constantly in search of new territories, a silent colonization that takes place thanks to the scattering of fertilized seeds. In this game, the strategy of sowing seeds to the wind—a metaphor that made Larousse dictionaries so successful (fig. 1)—is far from unique.

1 | *The Larousse dictionary's motto, "I'll sow it to the wind," illustrated by Eugène Grasset—or when culture spreads like dandelions' tufts.*

What is so striking in the first place is the incredible diversity of the very nature of these seeds, without any connection to the survival of the species appearing clearly. Since prehistoric times, humankind has benefited from the elegance of seeds. Their sizes and colors are so varied that it has made them choice elements in the making of necklaces or other jewelry.

Beyond their beauty, seeds are, for plants, indispensable actors in their life cycle. These are fertilized ovules that must first be protected in the mother plant. At maturity, scattering them must happen as efficiently as possible in order to ensure the sustainability of the species. Evolution has adopted several particularly clever mechanisms for scattering seeds.

Hidden Travelers

For a seed, the simplest strategy is certainly to drop at the foot of its plant, although this only ensures short-distance scattering. This is probably why lazy seeds subsequently use a different means of transport depending on the species.

Birds swallow these hidden travelers such as mistletoe seeds and then eliminate them after flying for a bit. Whether it is lightweight, like dandelion achenes that act like parachutes, or winged like maple samaras, a seed soon drifts away from home with the wind, unless it floats, like a coconut, to be tossed and washed away by ocean currents. Turning themselves over to randomness, a few lucky seeds will eventually find fertile ground for germination.

Plant Muscles

Some plants literally eject their seeds. But how, without any muscles, do plants project their seeds? Despite their placid demeanor, plants are actually capable of movement, sometimes even violent movement, often driven by the water carried through their tissues. These processes are usually slow, since water must travel through porous membranes or between cells to expand or shrink tissues and change the overall shape of the plant. This is the case of the pine cone: we have seen how it opens and closes as the humidity in its environment changes [THE CHOREOGRAPHY OF PINE CONES]. Even if the resulting movements are slow, one can use them to arm awesome grain-throwers, sorts of explosives that propel the seeds at great distance as we will see.

Plant Catapults

Ferns use what can only be called a catapult (fig. 2). A fern's spores are hidden under its leaves (which is why ferns are called *cryptogams*) in small sacs (or *sporangia*) about 1 millimeter in size. On their surface, they carry a sort of reinforced arc, consisting of a dozen cells full of water. Under the effect of heat, the water tends to evaporate, which leads to a drop in pressure: the arc gradually curls up, and the catapult is loaded.

When the pressure is low enough, a vapor bubble suddenly appears, and the cells abruptly resume their original shape. The slowly stored elastic energy is released in less than one ten-thousandth of a second, resulting in spore ejection at a speed of more than 10 meters per second. Slowed down by the air, they travel only a few centimeters. It is not much, but enough for the seeds to move away from the leaves and join the air currents that will carry them over greater distances.

There exist other modes of explosive scattering. This is the case of some legumes whose peas are contained in pods made of two joined shells. At maturity, they dry in the sun. The outer and inner materials of the pod then contract unevenly, or in crossed directions, which leads to the accumulation of mechanical stresses. Beyond a certain threshold, the seal between the walls fails, and the pod completely splits. This rapid ejection is triggered when the sac that contains the legume's seeds ruptures, which ejects the seeds as far as 1 meter from the plant.

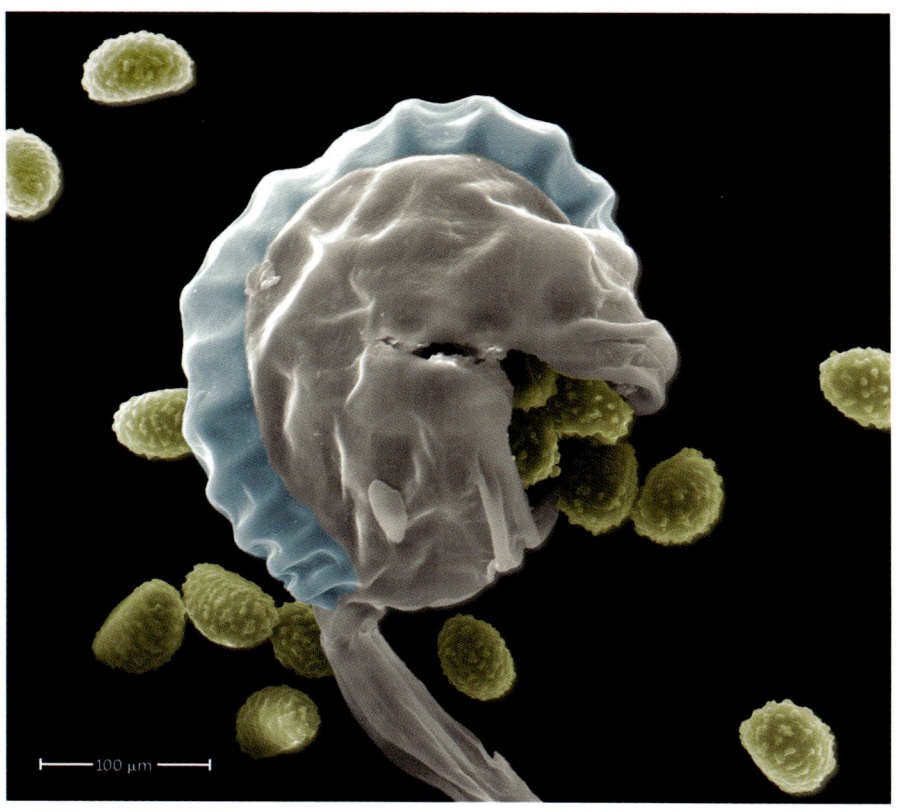

100 μm

2 | *This fern sporangium seen under an electron microscope contains spores (colorized green). When drying, the outer ring (colorized blue) changes its curvature by loading a catapult system.*

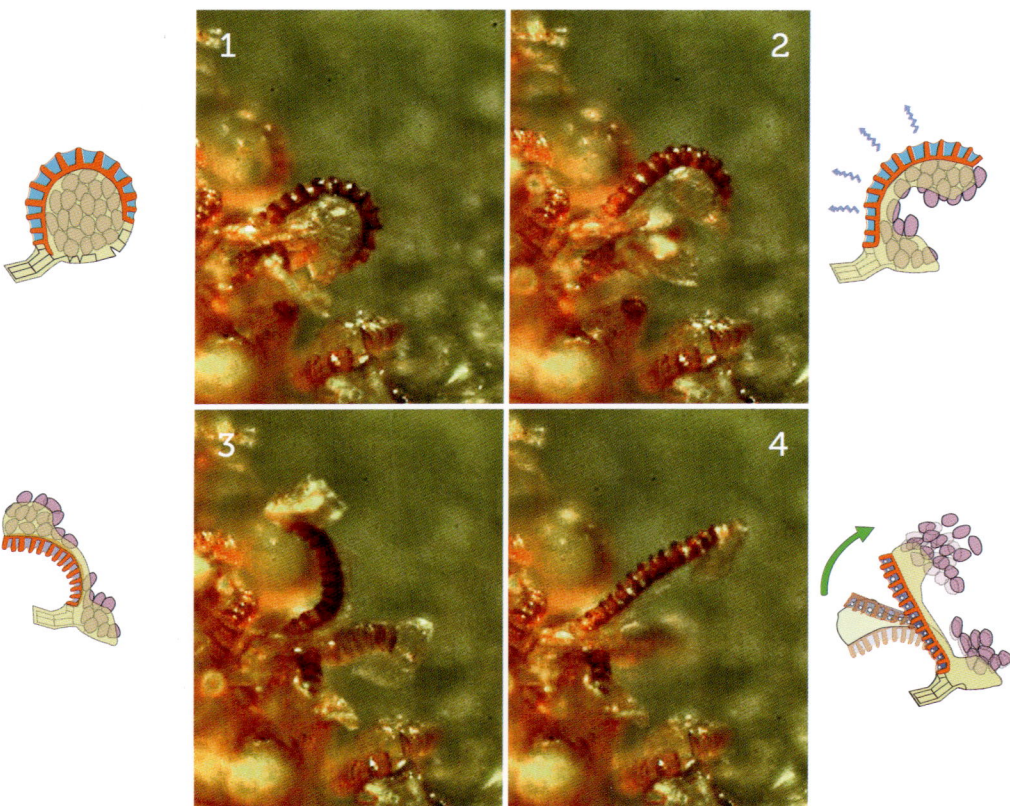

3   Fern leaves hide true catapults. When drying, the cells that make up an arched structure on the surface of a sporangium (the sac that contains the fern's spores) tend to deploy this structure and arm a kind of trigger. Relaxation occurs thanks to the violent boiling of the water that is still in the cells. Then the spores are ejected.

The Liveliness of Carnivorous Plants

Sudden movements don't exist just to eject seeds. You may have already encountered a Venus flytrap, a carnivorous plant, at your florist. Its leaves form a formidable trap. Each of them is in the shape of two lobes curved outward when the trap is armed. But all an imprudent insect need do is venture onto one of its jaws to activate one of the hairs that cover the surface and for both hulls to close on the insect in a fraction of a second, just like a wolf trap.

How can a Venus flytrap, without any muscles, be so quick? You've probably seen these small toys that were so popular in the 1980s: rubber poppers. They are rubber shells, armed by inverting their curvature, before

4   *A Venus flytrap in the open, waiting position (left) suddenly slams its two lobes shut by inverting their curvature (right).*

dropping them to the ground. At impact, the shell very quickly reverts to its original shape and jumps more than a meter high.

A comparable mechanism is at work in the Venus flytrap. In fact, because of their shape, the leaves' lobes lie at the limit of a mechanically unstable state [BEAMS BEND BUT DO NOT BREAK]. Stimulating the hairs triggers the small flick that makes them return to their original shape. It will then take several days for the trap to reload—time enough for water to flow through the cell walls and return the lobes to their unstable state. This kind of shell instability is also the solution bladderworts adopt—those small aquatic algae in the form of bladders that suddenly gobble up small crustaceans.

Other plants use speedy movement for defense purposes. The "sensitive plant," or *Mimosa pudica*, is so named because it almost instantaneously recoils when it is touched. A jolt from the wind, rain, or touch triggers an electrical signal that causes the water contained in the cells located at the base of the leaflets to evacuate to neighboring tissues. This local variation in the turgor pressure causes folding. The sensitive plant has even developed a remarkable, graduated response to aggression: the enfolding of its leaves spreads all the farther with the violence of the impact. The feats of plants will not cease to amaze us.

EXPERIMENT

To imitate a Venus flytrap you can arm a rubber popper slowly by hand. When you drop it flat on the floor, level, it resumes its hemispherical shape in a tenth of a second (1, the jump). The elastic energy released allows it to jump up 1 meter high. That's movement without muscle!

And what if you tried a slower experiment and returned to childhood for a moment to play with samaras? Endowed with wings, these maple seeds fall in a swirl. Recreate their fall by cutting small, elongated rectangles (2, the fall) out of a sheet of paper. Use a drop of glue to load some of them on a corner, a piece of Scotch Tape, or a bit of modeling clay, and then throw them. Depending on its mass and size, and even its shape, such a seed model rotates at various speeds and drifts with the wind.

Materials

1. The jump

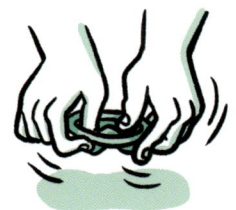

POING!!!

2. The fall

QUAKING BOWS

From Pernambuco wood from Brazil to horsehair from Mongolia, no material seems too luxurious for the bow, such an indispensable companion of stringed instruments—violins, violas, cellos, and double basses. But why such exotic materials? What dialogue emerges from this string and bow duet? You'll see that it is a small mechanical miracle all of its own, based on mini-quakes.

For any classical music lover, Johann Sebastian Bach's Cello Suites are a sort of ultimate accomplishment because of the power and variety of emotions that spring from the instrument. It sometimes seems to cry or sing like a human voice—composer André Jolivet (1905–1974) described the cello as a "demanding tenor." But what is the mystery of this evocative warmth all about?

1	*The bow's hair and the cello's string, which the bow rubs, work in tandem to produce sound.*

To find out, let's look at how notes are produced. A harpsichord or a guitar emits a sound when one of its strings is plucked and vibrates. The mechanism for producing sound in a piano is identical, the only difference being that its strings oscillate when hammers strike them. On the other hand, a stringed instrument introduces a much richer dialogue between the strings and the bow: the notes are produced by tiny seismic events when they touch one another. This strange effect makes the bow the secret weapon of stringed instruments; it is in a way an extension of the musician's arm and hand. It's quite simple really: "A violin is its bow," in the words of the Italian composer Giovanni Battista Viotti (1755–1824).

A Pernambuco Bow

Bows acquired their present shape and their nobility during the nineteenth century, first in the studios of Tourte, later of Persoit, and then of Peccatte: three generations of French bow makers who left their mark on the history of bow making as much as the Stradivari and Guarneri families did on violin making. Observe the stick of the bow: it exhibits a beautiful, concave profile that thins slightly from the heel. This profile used to be convex, like an archery bow, in the era of Italy's sixteenth-century musicians. Concavity allows the stick to be closer to the hair. The hair's tension then induces a weaker lever arm on the wood, which limits the bending of the wood. It is thus possible, given the same tension, to use a slender, light, and manageable stick.

The stick is carved of Pernambuco wood, a rare and protected species from Brazil's Northeast. This wood is particularly dense and has no knots; it is a combination of stiffness and flexibility that allows musicians to adjust the pressure they exert on the strings. Equally precious are the ivory, ebony, and mother-of-pearl present on the frog and the button that stretches the hair (a screw-nut system). The use of noble and exotic materials helps make the bow a valuable tool, sometimes as expensive as the instrument itself. This contrasts with Renaissance instruments, which used local resources as do some contemporary artisans (fig. 2).

The Hair That Came In from the Cold

What's with hair? Hair and the modern bow are quite a story. Whereas instrumentalists have gut, steel, and synthetic fibers to use as strings, choosing hair is a more restricted affair. Violin hair is almost always horse-tail hair. And not just any horse-tail hair, mind you. The most noble hair, because it's the most homogeneous in its composition, comes from the slow-growing stallions that live in the steppes of Siberia and Mongolia. Their horsehair is therefore a rather expensive material which, surprisingly, has never been displaced by fibers of another nature or contemporary composite materials.

2   *This viola da gamba bow was made by bow maker Coen Engelhard from the Ariège, a southern area of France along the Spanish border, using local resources: acacia wood and horsehair from the tail of the white horse in the background.*

Horsehair thus has specific qualities that are difficult to mimic. To determine what they are, we examined horsehair using a scanning electron microscope. We noted that there exist surface textures shaped like scales (fig. 3). This pattern is characteristic of natural hair. That is what, for example, explains why sheep's wool becomes felt when it is rubbed: when entangled, the scales transform the hair into a continuous fabric.

So, what role do these scales play in the production of sound? It is difficult to provide a conclusive answer because to understand the bow hair's interaction with the strings, you have to take into account a third essential element to the production of sound: rosin. Rosin is a wax made from pine sap that musicians regularly apply to bow hair. When the violin is being played, friction warms it up locally, which makes it less viscous at the moment when the rosin meets the string. Sometimes it even

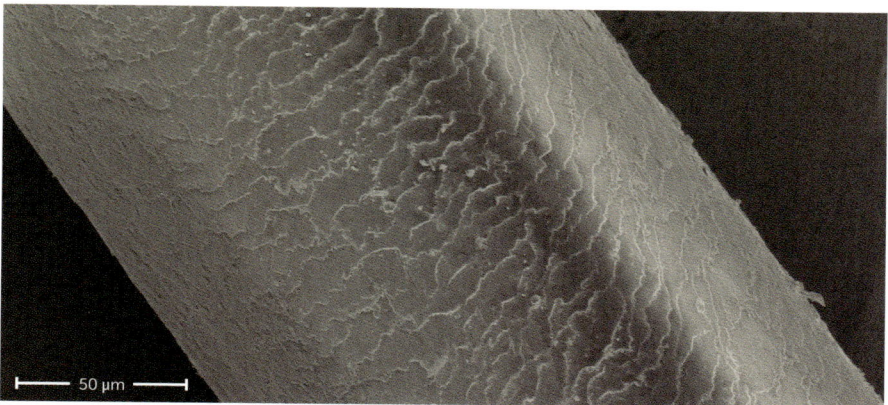

50 µm

3   *White bow hair visualized through a scanning electron microscope (its diameter is one-tenth of a millimeter). The scales (also present on our hair) all oriented in the same direction allow one to anchor the rosin more firmly on the horsehair. It seems unlikely, however, that scales play a direct role in sound production where the effects of two-way friction are the same.*

melts locally. It is the complex sliding of the bow on the strings that directly causes this heating-up mechanism.

The Stick-Slip Phenomenon

While German physicist Hermann von Helmholtz (1821–1894) is best known for his major contributions to physics and mechanics, he was also the first to describe the mechanism for producing sound with a bowed string, among other top-class works in the acoustics of musical instruments. It consists of alternating phases of static rubbing, when the string sticks to the bow hair and moves with it while stretching (this where rosin plays an irreplaceable role), and phases of dynamic friction when the string skids and vibrates.

To understand these two notions, think about what happens when you try to push a piece of furniture on the floor. As long as you don't push hard enough, it won't move: the force of friction counterbalances the force you're exerting, and the piece of furniture remains stuck, so to speak, to the floor. This force is called *static*. But, beyond a certain push, the piece of furniture will slide: that is *dynamic* friction at work. You will notice that if you continue to push, the effort you provide is lower than in the first phase when you failed to achieve your goal.

This is what happens with the string a violin bow stretches: as long as the tension on the string is not great, the bow grabs the string and deforms with it. Then the string is released, and it slides freely when it exceeds the limit of static friction, until it sticks to the bow again, when its deflection becomes weak enough, and it all starts all over (fig. 4). This sequence constitutes a cycle of movement, a *period*—the time between two sliding phases—that defines the note the instrument plays.

You might also be familiar with another example of sound-producing vibrations: the friction of a slightly wet finger along the edge of a crystal glass (fig. 5). All things considered, it is like bowing the strings and getting sound from the box of the violin.

On a larger scale, a similar phenomenon takes place in the earth's crust and results in violent earthquakes. Two tectonic plates in contact along a fault exert a shear on each other. Because of the continental drift, that shear keeps increasing; when it exceeds a static friction threshold, the plates begin to slip, causing an earthquake. In a way, studying the music of a violin, a viola, a cello, or a double bass amounts to better understanding how earthquakes are triggered.

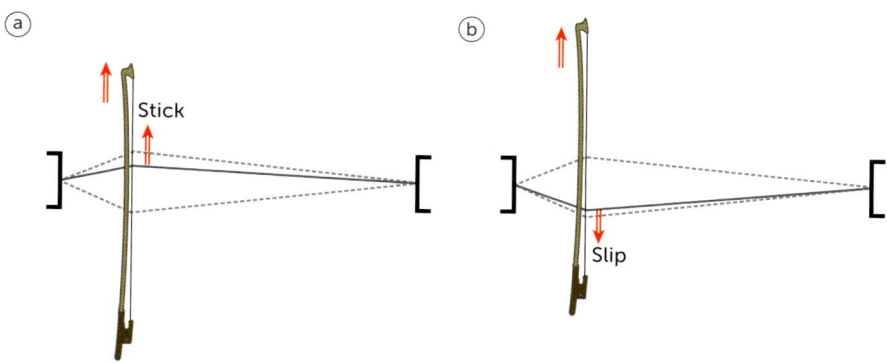

<table>
<tr><td>4</td><td>Here are the two phases of the relative movement of the string with respect to the bow's movement, which is continuous at this scale. (a) In the first phase, the string sticks to the bow hair. (b) When the string's tension is sufficient, the string releases and slides on the horsehair, until the string's deflection is again weak enough to stick to the bow again.</td></tr>
</table>

5 | *Singing glasses are another musical illustration of the stick-slip phenomenon. A musician first wets his finger in order to refine the properties of adhesion. The finger's friction on the glass involves a sequence of catching and slipping that make the glass vibrate. By lining up an ensemble of glasses more or less filled to adjust the note a glass produces, you can play a melody.*

EXPERIMENT

It is possible to see this alternating catching and slipping at our scale. To do this, attach the end of a long and flexible rubber band to a small wood block placed on a horizontal surface. Then pull the rubber band at a constant speed. To make the connection with a violin, think of the fixed plane as the bow and the rubber-band-mass unit as the string. The phenomenon is then seen from the perspective of the bow, from which the point of attachment of the string moves along at a constant speed. As long as the elastic is not stretched too tight (1), the force exerted on the block is not enough to set it in motion: that is analogous to the "stuck" phase of the bow.

When the rubber band is stretched and taut enough (2), the block suddenly starts to move (3): that is the "slipping" phase, during which the elastic loosens up. The block then stops, and the cycle begins again (4), (5), (6)—as happens when the periodic actuation of the string is induced by the bow's horsehair.

This model experiment is often used to demonstrate the mechanism of earthquakes. While stick-slip repetition happens quickly in this experiment, it takes place very slowly in earthquakes, whose period of occurrence is not only centuries-long sometimes but also irregular. Indeed, in the case of earthquakes, the geological faults are extensive and are therefore likely to have many points of friction. Some may slip and then suddenly drag others into an avalanche, causing one of those earthquakes of great magnitude, which remain unpredictable at the present time.

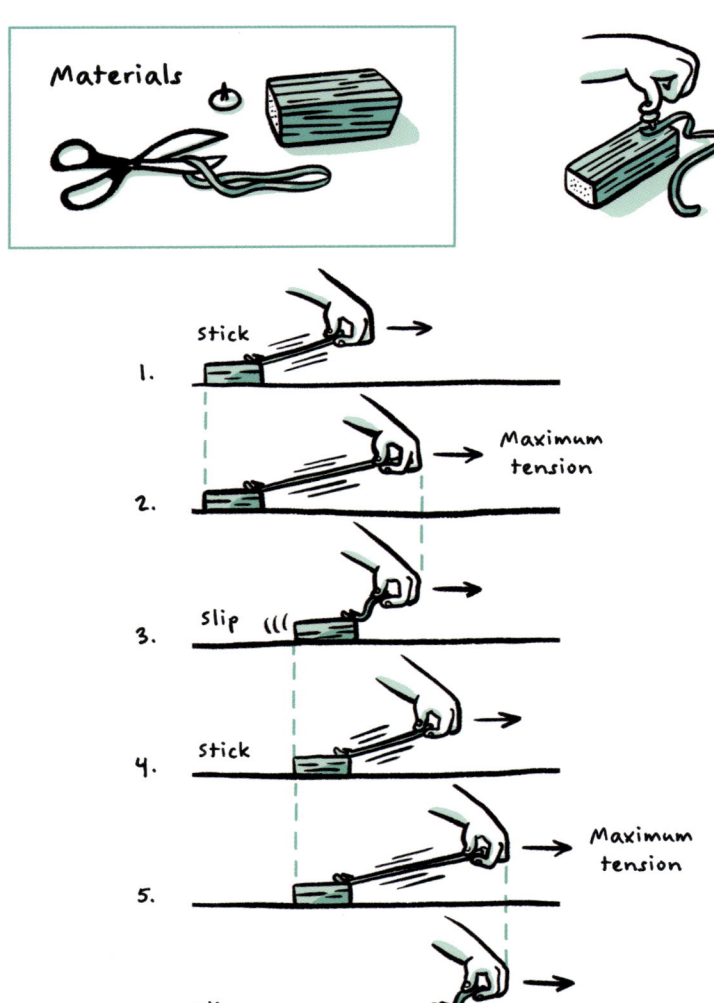

Materials

Stick

1.

2. Maximum tension

3. Slip (((

4. Stick

5. Maximum tension

6. Slip (((

RESTLESS GRAINS OF SAND

Contemplating a simple pile of sand makes for a beautiful science lesson, provided you see a solid that sometimes flows like a liquid. This is how physicists have come to understand why sand sculpts regular and universal shapes—as a result of the friction among the grains.

Capturing time using an *hourglass*. As early as the thirteenth century, this instrument helped seafarers navigate at sea and became so commonplace that it symbolized the inexorable course of life in artworks, alongside some skull or other *vanity*. Today, it is no more than a decorative element, and it takes an artist's or a physicist's eye to reveal its wonders.

1	This two-dimensional hourglass is the work of artist Jean-Bernard Métais: Temps imparti [Time Allocated], 2000. Thin and rectangular compartments are separated by a solid, perforated, horizontal surface. The flow creates triangular mounds in the lower compartment, and almost symmetrical craters on the upper level.

A Universal Angle

Let's take a closer look at the lower compartment of an hourglass and the sand pile that forms as the sand flows. In truth, the word *pile* is misleading because it connotes disorder and abandon, or stuff one might get rid of. But, even in its most ordinary form (a mound of dirt left by diggers), one detects a regular cone. Its appearance is not random. If sand is poured on an already formed pile, the added sand enlarges the cone, while maintaining its slope, its *angle of repose*, at a constant value. By slowly depositing grains of sand, one can make the pile steeper, up to the limit of its stability. But disturb the surface just enough, and an avalanche will form, and the slope will revert to its angle of repose. Researchers have shown that this angle depends on the nature of the grains: it is greater for rough or irregular grains (shell sand, chippings, gravel). But, curiously, it depends little on their size.

What physical ingredients go into this mysterious angle of repose? Unlike wet sand [THE SECRET OF SANDCASTLES], the grains of sand in a pile do not cohere. But grains have weight; they rub against each other and block each other, as long as the slope has not reached its angle of equilibrium. This problem is like that of a brick set on a tilted plane: the brick won't move as long as a threshold angle of static friction has not been reached—which depends on the nature of the surface of the materials that are in contact [QUAKING BOWS].

Trapping Prey

There is an insect—an insect with an innate sense of physics—that has been able to take advantage of this angle of repose: the antlion. Its larva captures its prey by boring a hole in the sand, which it penetrates (fig. 2). It is a formidable trap for ants venturing into the funnel: the ant's desperate efforts to return to solid ground create an avalanche that brings the victim closer to its ruthless predator. All the antlion larva has to do is wait for its meal to drop straight into its mouth. Sometimes, it speeds up the poor ant's fall by catapulting sand to trigger an avalanche. There is, however, an optimal size of prey: an insect that is too light can climb up the fragile pile without disrupting it. And, conversely, a heavy beetle can dig steps to climb back up the slope.

2 *"The Ant-lion makes a slanting funnel in the sand. Its victim, the Ant, slides down the slant and is then stoned, from the bottom of the funnel, by the hunter, who turns his neck into a catapult." Jean-Henri Fabre, Fabre's Book of Insects (retold from Alexander Teixeira de Mattos' translation of Fabre's "Souvenirs Entomologiques" by Mrs. Rodolph Stawell). (Mineola, NY: Dover Publications, 1998), p. 108.*

Mysterious Barchans

In Japan, the patterns carefully drawn in the sand of Zen gardens invite us to meditate. For anyone who has experienced it, a sea of dunes that sweeps across the Sahara has the same calming power, though at a different scale. It is no doubt because dunes are essentially very big mounds of sand. As such they have the same attributes as piles of sand, including an angle of repose. Contrary to appearances, a dune changes shape and moves around continuously with the wind, which constantly carries grains of sand that it deposits on dunes.

Among the different morphologies of sand dunes, crescent-shaped barchans are among the most elegant (fig. 3). They appear under the effect of a prevailing wind in the vicinity of a coast swept by trade winds; or in

3 | *A barchan dune shaped by a wind coming from the left. The wrinkles perpendicular to the wind on its upper side are similar to those that are shaped on a beach at low tide.*

the Sahara Desert, pushed by the dry Harmattan wind out of the northeast. The winds carry grains of sand that move by jumping (or *saltation*) to the dune that is forming and climb the gentle slope of the part exposed to the wind. At the crest, they tumble down in small avalanches on the sloping and concave side (at the angle of repose). Avalanche after avalanche, a barchan can thus move several meters each year, and the smaller it is the faster it moves. Marco Polo was the first to describe the sound poetically called a "dune song," which accompanies these avalanches and whose secret physicists have recently unveiled. This sound occurs when the flowing surface layer of grains vibrates, like a drum membrane, on the fixed lower bed.

Dunes can be found in many different environments: at the Great Pilat dune along the Atlantic coast of France; in desert fields; in the high mountain plateau of Great Sand Dune Park in southern Colorado (where the tallest dune is 210 meters high) molded by western winds.

Dunes on Mars?

In the early 2000s, space probes sent to Mars showed that the earth was not alone in having dunes—one could see superb ones on the red planet (fig. 4). Their shapes are identical to their Saharan cousins', but they are ten times bigger. In contrast, underwater dunes are a hundred times smaller than their counterparts on solid ground. This universality of dunes is a blessing for physicists: their comparative study confirmed the models that describe the formation of dunes on Earth.

There is a critical size for a dune to develop. This size depends on the ratio of the density of the grains to that of the fluid in which they move. Underwater dunes, which occur in a fluid a thousand times denser than air, rise to minimal heights a thousand times lower than those of the desert, whereas those formed in the rarefied atmosphere of Mars are thirty times larger than those on Earth.

4 | *The prevailing wind from the left accumulates sand upstream of this Martian dune, which is the counterpart of Aeolian dunes. Avalanches occur along the steep downstream side, which contributes to the dune's advancing. There are also regular wrinkles on the dune's dome, as in the desert.*

EXPERIMENT

In order to observe the triggering of an avalanche, fill one third of an empty, clear CD jewel case at its center with dry sand (1); then seal it with adhesive tape (2). Rotate it gradually on a vertical plane. The initially horizontal bed of sand gradually slopes (3), (4), until an avalanche is triggered at a certain critical angle, the *avalanche angle* θ_m (5). The sand flows, the angle decreases and then stabilizes at an equilibrium value, the angle of repose θ_r (6).

The values of these angles depend on the nature of the grains and their surface condition, but the existence of these two thresholds is universal. You will be able to reproduce this experiment with a granular mix of different sizes and colors, which will display beautiful stratification effects.

Materials

1.

2.

3. $\theta = 0$

4. $\theta < \theta_m$

5. $\theta = \theta_m$

6. $\theta > \theta_m$ θ_r

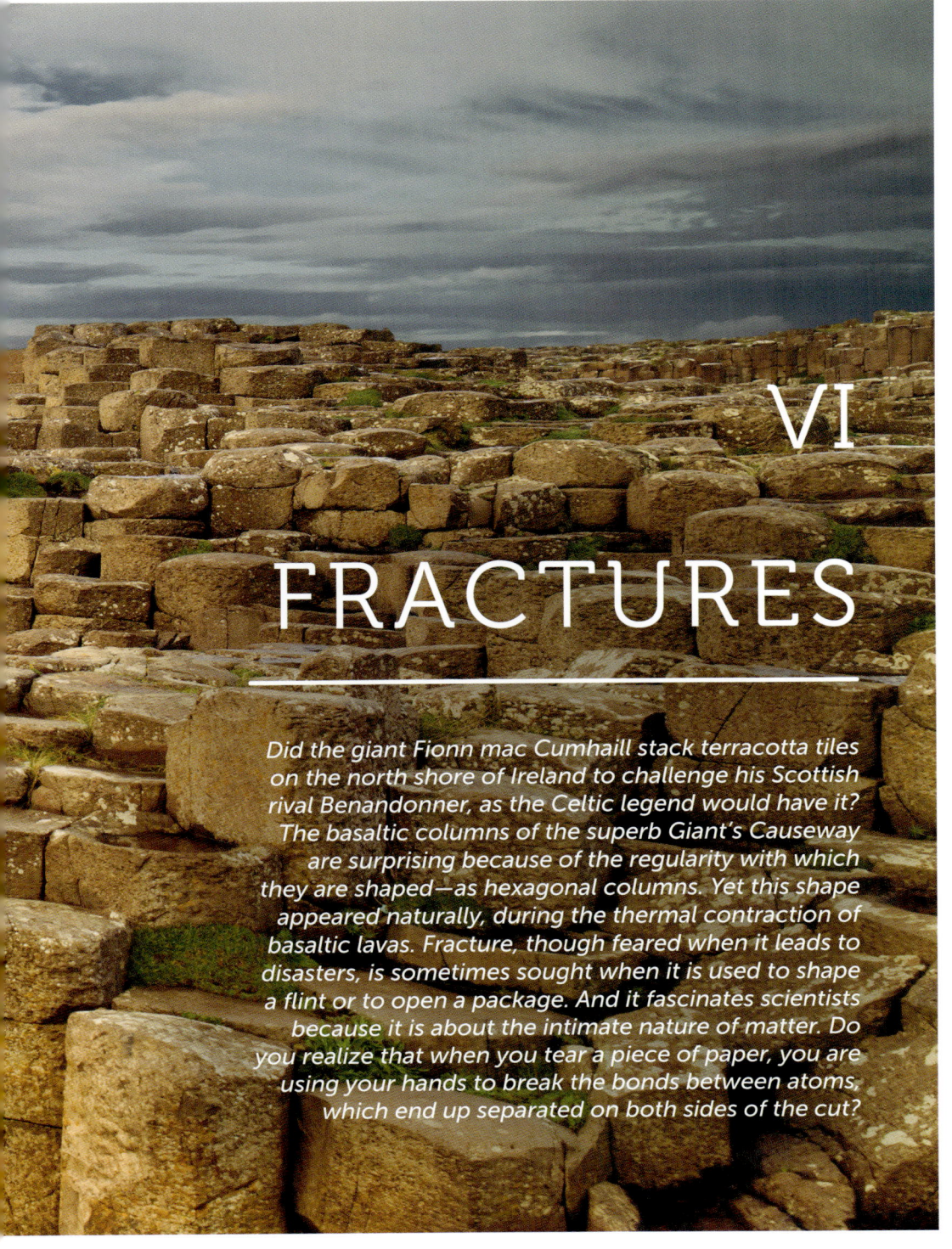

VI
FRACTURES

Did the giant Fionn mac Cumhaill stack terracotta tiles on the north shore of Ireland to challenge his Scottish rival Benandonner, as the Celtic legend would have it? The basaltic columns of the superb Giant's Causeway are surprising because of the regularity with which they are shaped—as hexagonal columns. Yet this shape appeared naturally, during the thermal contraction of basaltic lavas. Fracture, though feared when it leads to disasters, is sometimes sought when it is used to shape a flint or to open a package. And it fascinates scientists because it is about the intimate nature of matter. Do you realize that when you tear a piece of paper, you are using your hands to break the bonds between atoms, which end up separated on both sides of the cut?

PREHISTORIC GEMS

Carefully displayed in museum cases, prehistoric stone tools called bifaces arouse interest especially among prehistorians. Although ... if you look closely, the refinement and symmetry of the cut of some of them remind us of today's diamonds. Can we see in them any artistic intention?

Bifaces of stone are among the oldest vestiges left by our distant *Homo* ancestors, and yet it could well be that they are not tools, but gems. To prehistorians these bifaces indeed continue to harbor mysteries. Made by humans for at least 1.7 million years, they were first used to hunt and cut meat off animal carcasses, and much later to help in agricultural work. But some of these bifaces would feel right at home in a jewelry store because of their perfectly symmetrical pear shape like the Cullinan diamond of the English crown (fig. 4). The anthropologist Jean-Marie

1 | *A beautiful stone cut five hundred thousand years ago. This biface was discovered at Abbeville by Jacques Boucher de Perthes (1788–1868) and presented at the 1867 World Fair. It was named* Acheulean, *after the site of Saint-Acheul in the same area.*

Le Tensorer sees in these prehistoric gems the birth of harmony and the early fruits of artistic creation. His interpretation is based on the fact that these objects have none of the microscopic defects brought about by typical use. These remarkable bifaces were probably not utilitarian and would have been used as adornments.

Metallic Clinking

How did our ancestors get such beautiful bifaces? First, they had to find a quality stone. The raw matter—flint—is found as nodules in limestone or clay soils. Both brittle and hard as glass, it breaks into blades and shards. A flint tool hardly wears with use. The same is true of modern ceramic knives [THE SAGA OF FUSING GRANULAR MATTER], which have a reputation for being resistant to wear but can break when simply dropped.

Flint nodules are formed by precipitation in sea water or a lake that is saturated with silica. Its chemical formula, SiO_2, which is part of the formula of many minerals, derives from changes in the silicate crystals of the continental crust, oceanic volcanoes, or from the decomposing siliceous exoskeletons of marine organisms. Ideally, the nodules ought not contain defects or impurities that would favor the stone's breaking up into a multitude of fragments. A simple way to check this is to hit the nodule and listen to the sound it produces: a metallic sound guarantees a homogeneous stone.

The Subtle Gesture of *Homo sapiens*

Can we recreate our ancestors' gestures, who transformed these nodules into bifaces? The traces of carving tools, found alongside flint waste on production sites, helped archaeologists like Jacques Pelegrin reconstruct carving methods (fig. 2). Our ancestors first used stones to knap flint nodules by striking them. But, over the ages, carving has gradually been refined: sculptors have replaced striking with stones by using harder impactors made of deer antlers, which provided more precise chipping on both sides of the flint. From the first coarse bifaces produced more than a million years ago, to the latest during the Neolithic era, a slow evolution would lead human beings to refine chipping operations in order to produce ever-sharper tools.

Indirect percussion on a hard wood precisely placed on top of flint would also constitute significant progress in this evolution towards the Neolithic. The chiseler chooses a fracture nucleation site and presses down on his tool. The elastic deformation induced by the sudden impact on the tool becomes manifest in a small bulge and the appearance of regular folds, signature signs of the strike's waves (fig. 3). The word *conchoid* for this smooth face refers to the open shape of a conch. The irony is that

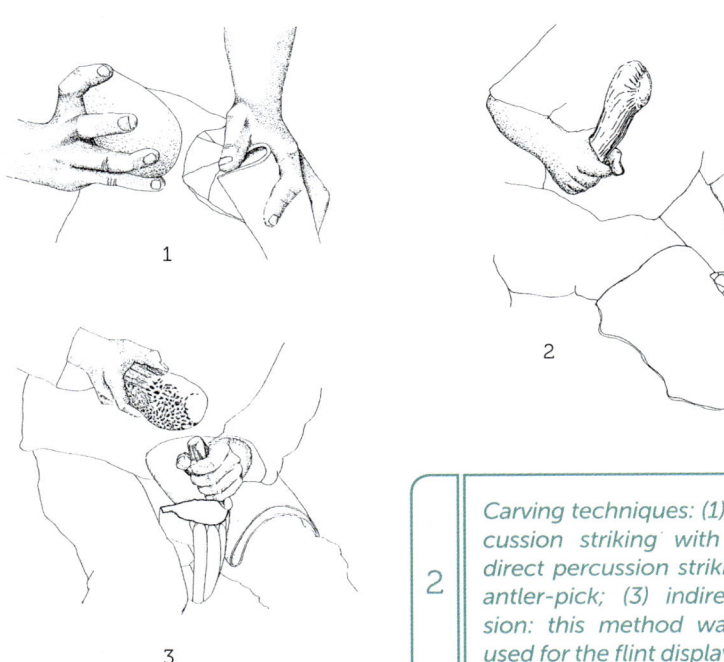

Carving techniques: (1) direct percussion striking with stone; (2) direct percussion striking with an antler-pick; (3) indirect percussion: this method was probably used for the flint displayed in fig. 3.

our ancestors empirically mastered these dynamic fracture problems, which are still open questions today in fundamental mechanics.

There is a last means of chiseling, by compressing a wooden rod, a kind of crutch, placed very precisely on one face of the previously prepped flint nucleus. Developed five thousand years ago, this technique has proved particularly effective in producing sharp blades. The strong compression of the rod causes it to bend slightly, storing a considerable amount of elastic energy. When the stone breaks, this energy is suddenly released and chips the flint at once. In addition to controlling the direction of fracture, this technique avoids the creation of impact waveforms. Its development has led to the industrial production of extremely thin blades, objects of much commercial exchange: in the French Jura, in Switzerland, and even in the Netherlands, parts were found that were produced several hundred kilometers away in Touraine.

Bifaces in Modern Times

Despite their analogous shapes, diamond cutting has little to do with flint knapping. Diamond cutters strive to enrich the play of light and the brilliance of the final gem. Their task is made more complicated because diamonds are the hardest material on Earth. Only a tool covered in diamond powder, or friction against another rough diamond (or *boart*) of poorer aesthetic quality, can saw, wear down, and polish the facets once they have been cut by impact in the direction of certain crystalline planes. The facets' regular geometry is such that the light that penetrates is reflected and dispersed several times on the inner facets of the diamond, because of its very high refractive index (2.55 vs. 1.5 for glass). It's hard work: a classic cut, the brilliant cut diamond, has fifty-seven facets.

The initial cuts aim at making several gems from the same raw sample while minimizing any loss of material. From the enormous, exceptional, 620-gram (approximately 3,100-carat) Cullinan rough diamond found in the Transvaal, South Africa, in 1905 and donated to the King of England, an Antwerp-based diamond cutter was able to create nine large

3 *Side by side: a piece of flint (or nucleus) of the "pound of butter" type— because it evokes the butter formerly pressed in wooden molds—and a large blade that was chipped from it (about 2500 BCE, Abilly, Museum of Prehistory of Grand Pressigny). The initial strike was on the left facet of the block, its heel. The nucleus is identified by the initial rim on the left (blue arrow) and a series of small periodic waves toward its front end (green arrow).*

diamonds and nearly one hundred brilliant cut diamonds, some of which are smaller than 1 millimeter in diameter. Opticians, geometers, mechanical engineers, artisans, and artist jewelers all perform delicate work. In France their activity is well represented by stone cutters in the Haut-Jura region (near the Swiss border), where stone cutting dates back to the sixteenth century.

EXPERIMENT

You can safely imitate a stone cutter's gesture—using chocolate. Gently melt a bar of baking chocolate, stirring it in order to homogenize it (1). Pour the melted chocolate into a cylindrical container and place it in the freezer (2), (3). This will create a chocolate nodule (4), which you can chip at using a wood chisel and a mallet (5).

Before tasting the strange chips produced, take the time to look closely at the smooth facets and the fracture lines.

4   *The pear shape of diamonds, like the Cullinan 1 of England's 74-facet tiara, is reminiscent of pre-historic cut stones.*

Materials

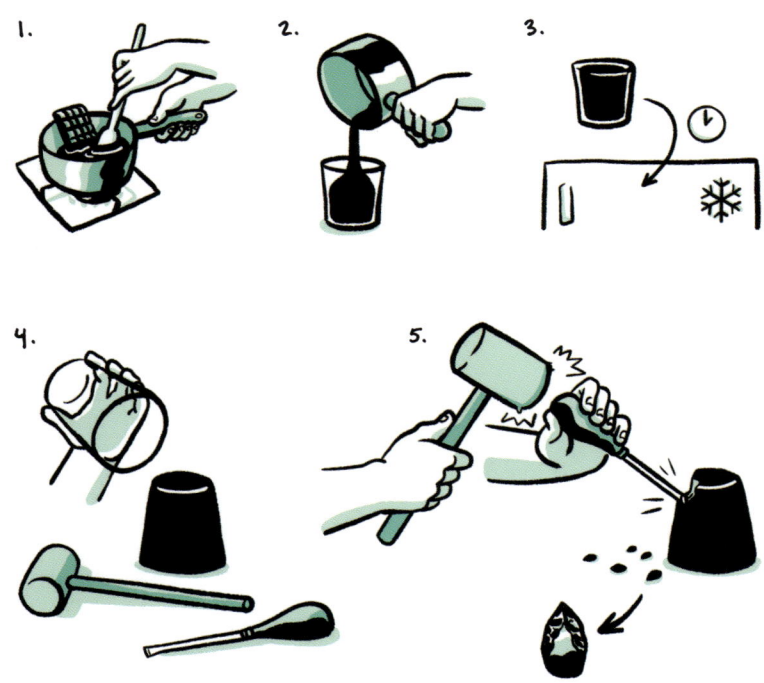

269

POINTED
TEARS

Why would rips and tears in a poster interest both artists and physicists? These rips reveal great collective artworks and also help understand how cracks propagate through the fuselage of an airplane.

The work of contemporary artist Jacques Villeglé at first evokes the flight of white and blue birds with tapered wings. But look more closely: what appears as a collage of fragments of colored sheets is actually a piece of fence covered with layers of posters. Passersby tore off pieces of these posters, revealing the contrasting colors of the underlying layers. Once these overlays are framed and hung on the wall of a museum, the urban rips leave behind anecdotal degradation to enter the realm of beauty.

1 | L'Éclatement des Célestins, *Jacques Villeglé, August 17, 1964.*

This abstract painting without paint is for Villeglé the collective work of so many anonymous artists. For the physicist, a hand's gesture, which reveals unexpected patterns through the sharp rip shapes, illuminates the way such thin sheets tear. Remember the last time you tried to remove a piece of Scotch Tape from its roll? It's often a frustrating experience. If you fail to take off the strip across its entire width, it is almost impossible to correct your move: the partial strip narrows and adopts precisely the same triangular shape, which is even more difficult to detach.

Midflight Rips

If you pay attention, you'll see such sharp, pointed shapes in all kinds of situations with thin film tears, whose thickness is much smaller than their length and width. Very flexible tissue paper of this kind is commonly used in packaging—Christmas gift wrapping, for example.

2 | *Look closely: peeling a blanched tomato also creates sharp flaps.*

On a larger scale, sheet metal is also comparable to thin films. Aircraft fuselages can be seen as basic structures wrapped in a relatively thin aluminum sheet, and fuselage tears will have the same profile as sheet metal tears in a breakage situation.

An Aloha Airlines Boeing 737 unfortunately illustrated this phenomenon in 1988: on its way to Honolulu, the upper part of the cabin was sheared off after depressurizing at an altitude of more than seven thousand meters—and note that fortunately the pilots did manage to land their mutilated apparatus. And, like adhesive tape or a torn sheet of paper, the damaged fuselage presented a series of sharp tears. At another scale, you can see the same type of tear when you peel a roasted pepper or a blanched tomato (fig. 2).

What Makes the Tears Pointed?

To what do we owe the universality of these pointed tears? The thinness of the sheet is key. To understand it, let's take a look at a tomato: when you pull its skin to peel it, it folds markedly where it separates from the fruit. A general property of thin films is that they are much easier to fold than to stretch (try it with a sheet of paper). The tear occurs only when the skin becomes too curved along this fold. Indeed, the cracks propagate by reducing deformations, which can be accomplished by narrowing the width of the curved area. That is why the area that is folded naturally tends to diminish as one peels a tomato. When pulling on a flap, tears propagate by focusing: the flap adopts the shape of a point.

A Nobel Prize Thanks to Scotch Tape

The same tear mechanisms apply at a microscopic level. Do you know what the thinnest leaves are that scientists have been able to come up with? The answer is graphene sheets, which consist of a single layer of carbon atoms arranged as a hexagon—a configuration reminiscent of terracotta kitchen floor tiles. This amazing material was discovered in 2004 thanks to graphene's ability to peel easily.

It all started at Manchester University, during an experiment done at a Friday evening meeting of researchers, together to test new, somewhat crazy, and fun ideas. That evening, Andre Geim and Konstantin Novoselov observed that it is possible to peel away very thin layers from solid graphite (a controlled version of the pencil lead that leaves behind a black graphite film on a sheet of paper). How? Simply by *sticking* a strip of tape to it and peeling it off.

By gradually removing material the same way from the already thin layer of this material, they made the material thinner and thinner, until monatomic layers were left on the adhesive. Incredible as it may seem, these researchers were able to isolate layers as thin as an atom with simple adhesive tape. It turned out that, often, during these successive peelings, the film was partially torn, so that nice pointed flaps also then appeared. Today graphene is considered a material of the future. Its exceptional electronic and mechanical properties make it possible to imagine many technological applications [FOLDING AND CRUMPLING PAPER BALLS]. In fair payback, this discovery earned its discoverers a Nobel Prize in 2010. The story of graphene can be summed up this way: how to get the greatest award thanks to a pencil, a roll of adhesive tape, and a few rips and tears!

3 | *Graphene self-tears: In this image, a layer of graphene was perforated by a diamond micropyramid that left a recessed area in the substrate (dark area) and triggered three pointed tears. But who is pulling on these few-millimeter-long flaps? It is the strong affinity graphene has for itself that tends to fold the flap to increase the self-contact area.*

EXPERIMENT

Stick some adhesive tape to a flat surface that is easy to clean (1) (a ceramic plate, for example, that can be scraped without damaging it). Make two small cuts in the tape (2), and then pull on the flap you've made (3): the shred will inevitably be triangular (4). If you carry out the same experiment changing the distance between the initial notches, you'll see that the length of the flap changes but that the angle of the triangle remains the same, provided you draw the flap at approximately same speed. A researcher would call this a "robust" result.

The angle of the tip depends slightly on the speed at which you pull the tape. On the one hand, if you pull it quickly, the effective adhesion of the tape is greater and leads to a more obtuse triangle. On the other hand, by pulling slowly you will get a sharper or more acute tip: the greater the adhesion, the harder you have to pull on the flap and the more the fold bends, thus more effectively pulling tears inward. You can also get pointy tears with plastic packaging film lying flat on a surface, in the absence of any sticking. In this case, the flap is not exactly triangular.

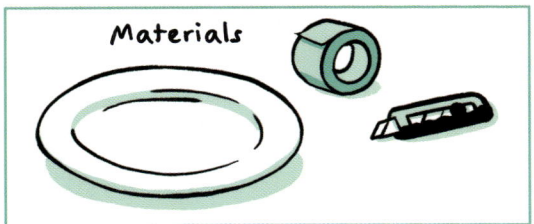

Materials

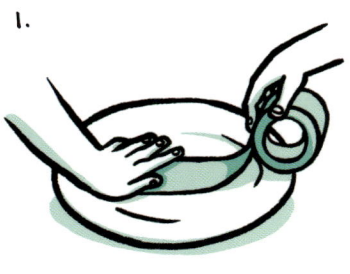

1.

2.

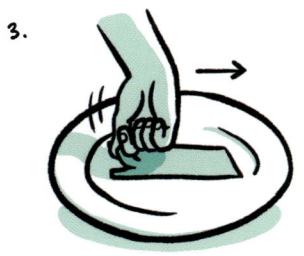

3.

4.

THE GEOMETRY
OF LETTUCE

How does one explain that a leaf of salad bears the same undulating folds as the body of some sea slugs or a torn sheet of plastic? What physical ingredients hide behind such similarities?

They do not belong to the same world, yet they resemble each other! On the one hand, a lettuce, one of the many examples of plants with wavy leaves; on the other, a sea slug, *Elysia crispata*, a distant cousin of the animal that sometimes invades our gardens. This mollusk has developed a

> **1** *Lettuce or sea slug? What does a mollusk of the tropical waters,* Elysia crispata, *which is a few centimeters long, have in common with the lettuce in our markets?*

remarkable camouflage technique: it disguises itself as seaweed to avoid being eaten by carnivorous predators. Both a leaf of lettuce and this slug have wavy edges that evoke some Baroque ornament. The richness and variety of shapes in nature are certainly often astonishing, but how to explain such a convergence?

For a moment let's forget the living world. In order to understand where these undulations come from, researchers turned to rippled shapes in a totally different context: plastic bags, preferably thick ones. Try to rip one of them, a garbage bag, for instance: you will find that propagating an existing tear in this material requires effort—and that's a good thing!—unlike cookie-wrapping film, for example [POINTED TEARS].

Why are these bags so rip resistant? Simply because they are made of so-called *ductile* polymers—they have to be stretched before they rip. Thus, before being able to tear the material, you first have to expend a lot of energy to deform it irreversibly, a little bit like stretching chewing gum. In contrast, when a plate falls and breaks, hardly any energy is used to deform the material And we call that a *brittle fracture*. The same goes for the flint tools that our Paleolithic ancestors carved [PREHISTORIC GEMS]. This type of fracture often results in smooth, sharp edges. Now take a look at the edge of the tear of ductile film (fig. 2): in contrast, it presents a rippled, curled form, identical to that of the edge of a beet leaf or a sea slug. Might there be ingredients that are common to all three of these situations?

Let No One Ignorant of Geometry Enter Here

In order to get at the mechanisms that can generate such complex shapes, let's return to the experiment carried out on thick plastic bags. When you tear a ductile bag, it has to stretch considerably before it rips. In other words, the torn edge is strongly deformed and, moreover, permanently. Despite that, the areas of the bag far from that edge have not changed shape. A geometrical incompatibility thus appears among the different

> **2** *The edges of a torn plastic sheet (at the top) show undulations on several scales: inside large central waves that are centimeters thick, millimeter-thick waves can be observed, which themselves contain an even finer structure (at the bottom).*

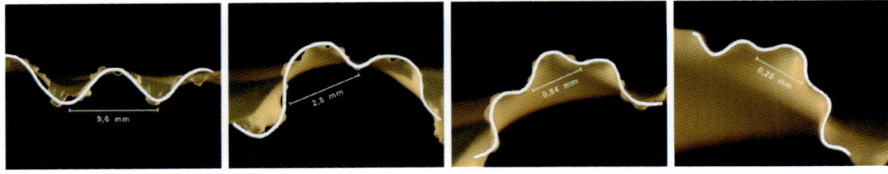

regions of the flap. How to reconcile this highly stretched edge and the unscathed part of the bag? Well, either by stretching the body of the bag or by squeezing the edge! Now, a simple experiment shows how much easier it is to crumple a sheet of paper, for example, than to stretch it: the compression very quickly leads to the sheet bending out of its plane. This, then, is at the root of the surprising convolutions of the torn edge: length incompatibility.

On closer examination, these convolutions are also astonishingly regular (fig. 2). Far from the messy folds of a crushed paper ball [FOLDING AND CRUMPLING PAPER BALLS], each curl seems to have been skillfully drawn in a pattern of waves and recreated at several scales. At a distance, they evoke a sort of high-sea swell, but looking a bit more closely, you'll make out waves within these waves, and then even smaller waves that you'll only notice if you get in even closer. It's dizzy-making! Mathematicians call this type of nested-doll structure *fractal*: it reproduces identical shapes at different scales. In the case of the torn plastic bag, you can make out up to five generations of ripples: the edge ceases to curl when the size of the wave is of the order of magnitude of the thickness of the sheet.

Managing Growth

Let's return to lettuce leaves and slugs. How could their edges become "too long" compared to their inner parts? Simply because they are living beings that grow! Curiously, the young leaves and juvenile slugs are rather smooth, but they gradually wrinkle as they grow. Everything seems to indicate that the edges have instructions to grow faster than the rest.

This phenomenon, although less intense, can be seen when certain flowers bloom (fig. 3). Initially smooth lily petals grow more actively near their edges than along their main vein, which makes the opening of the petals easier; then, in the last stages of blossoming, ripples appear. The undulations that finely chisel lettuce leaves might thus originate in a balance of physical forces combined with biological programming. How does a simple instruction generate shapes of incredible complexity?

Finally, the "simple" flat plant leaves, which are so common around us, are what ought to amaze us. Their flatness is indeed a sign of spatially uniform growth. Budding knitters are well aware of this, disappointed as they are with their early work when it's bumpy because of irregular stitches: it is more difficult to produce a flat surface than a wavy form.

Similarly, it is difficult to conceive such petals without a very precise control of flower growth, that is to say without regulatory strategies that resemble our hormonal systems. The implemented biological mechanisms, however, remain largely mysterious.

3 | A blooming lily. The ripples that appear at the end of the growth process (left petal) are caused by stronger growth at the edges.

EXPERIMENT

Using a ballpoint pen, draw evenly spaced (a few millimeters apart) parallel lines on a thick plastic bag (1), (2).

Pull out scissors and make a notch at the edge of the bag along its middle line (3), then tear it trying to follow this central line (4). The edge takes on a wavy form like the one described above.

Now try using the scissors to cut strips parallel to the torn edge, following the other lines. Then flatten them under a transparent plate (or a glass dish).

One might think that the only difference among these strips is their respective lengths, all the longer as you are near the strip's edge (5). Curiously, these strips curl spontaneously when we flatten them. This curvature is more pronounced near the torn edge. It reveals that one side of the flap is longer than the other. One can easily see that the surface that accommodates these different curvatures must be very complex.

Beyond a certain distance from the edge, the strips no longer curve: this indicates that this area did not undergo any irreversible deformation during the tearing.

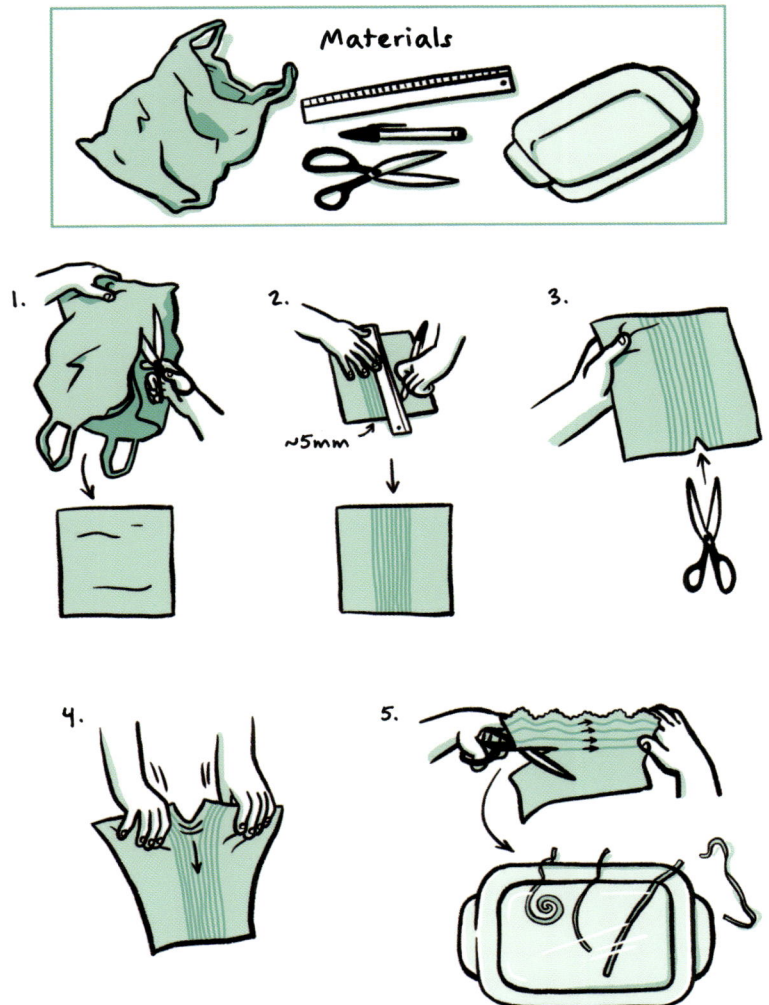

Materials

1.

2. ~5mm

3.

4.

5.

285

ELOQUENT CRAQUELURES

Only a researcher would be so interested in the microscopic defects of the Mona Lisa, that paragon of perfection. Dreaded or sought-after depending on the work, the cracks that occur on a thin, solid layer reveal the tensile stresses acting at the surface or deeper in. Similar mazes are found in surprising situations such as urban networks.

A detail is enough to recognize her unambiguously and also puts the influence of time in relief: a network of craquelures crisscrosses the canvas of the *Mona Lisa*. The nature of this geometric motif is nontrivial, since it allows specialists to tell its story from its creation to the present day. It is, moreover, thanks to these cracks—true digital fingerprints—that we were able to prove the authenticity of the painting that miraculously resurfaced in 1913 after having disappeared two years earlier.

1	*The Mona Lisa, a subject of multidisciplinary research, especially regarding its imperfections.*

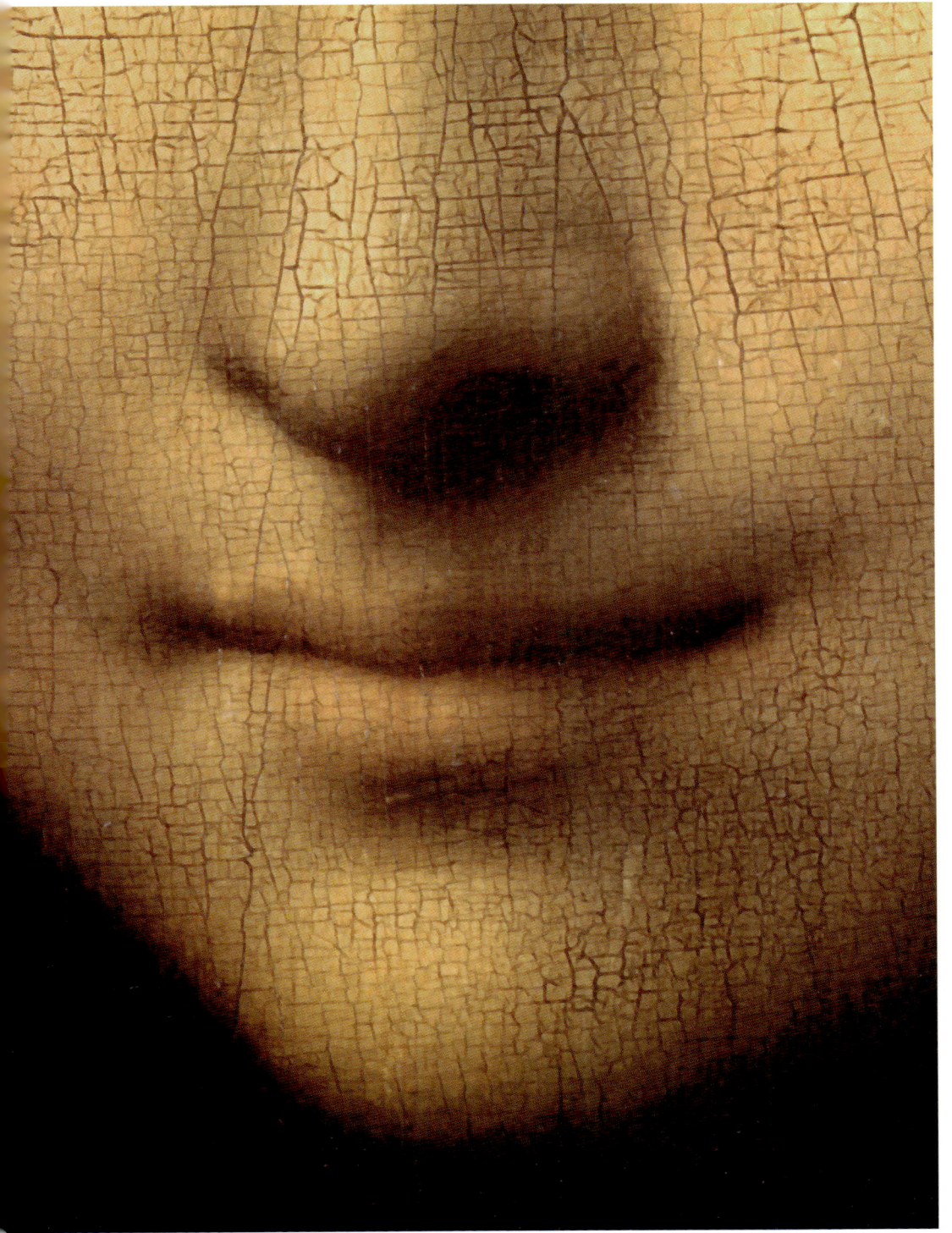

The cracks that scar the surface of the work are a valuable source of information. The deepest among them speak volumes about the deformations of the painting's poplar wood substrate which are the surface manifestation of cracks within. Networks of cracks are also born at the very heart of some paint layers, except for the thin surface deposits da Vinci made to soften the boundaries between colors and give a smoky impression (*sfumato* technique). In fact, all layers of paints and varnishes have their own cracks, hence the interest in studying these different levels of defects using appropriate lighting and medical imaging techniques.

Relieving Fractures?

But why do these cracks appear? They result from the strong tensions in the material. These stresses are transmitted from the substrate to the pictorial layer, or arise from within the thickness of the paint layer. Research-

2 | A cup covered with crackled enamel. Notice the hierarchical network of cracks, which connect at right angles. The cracks visible inside the cup, in which the tannins of the coffee are encrusted, appear more contrasted.

ers have identified the following mechanism: while the paint dries, some of the solvent evaporates so that the pigments come together, forming a solid film. That film tends to retract, but it is geometrically constrained by the substrate to which it adheres. Too strong a tension in the paint layer leads to it cracking or separating, which causes the mechanical stresses in the film to relax. Even after the paint dries and consolidates, cracks occur throughout the life of the work. But external aggressions are also numerous: fluctuating temperature and humidity, ultraviolet radiation, the air's oxygen, improper handling, or even bacteria.

Take a look at the elegant patterns that cracks draw in thin films, and you will notice that you can find them in everyday life: in the paint on our buildings, the varnish of our manufactured objects, or even the protective layer of solar panels. Most often, they form complex, interconnected, hierarchical networks. There are, however, also isolated cracks when stresses are low and the material has few defects to initiate a crack.

Cracks as Ornaments

Their diversity and aesthetic qualities also mean cracks are sought after to break up the monotony of a flat surface. Ceramicists thus apply processes that create such patterns, as in this cup (fig. 2). Glazes are applied to the surface of the object; after being fired, the enamel hardens as it cools, contracts, and cracks with a characteristic crackle. It is also possible, when the network of cracks has stabilized, to brush India ink on the surface, which penetrates the cracks. Rinsing the excess pigment in the water reveals the delicate, hierarchical network of fractures—your coffee, too, acts this way inside a cup.

Similar physical phenomena are at work in the production of certain paintings: a first layer is covered with a different color. Then you can see the colors of the lower layer appear in the fissures that emerge. This same principle is used in fancy nail polish to produce an effect of cracks that extend to the fingertips.

From Cracked Soil to Urban Networks

These networks of cracks are common in clayey soils when they dry up. The water initially present among the grains evaporates, and they tend to compact. The grains' compaction is thwarted by the substrate, giving rise to tension: a first crack will soon form. If this opening allows the material

Periods when streets were created

- pre-1000
- 1000–1300
- 1300–1450
- 1450–1600
- 1600–1700
- 1700–1790
- 1790–1850
- current roadways

3 On this map of Paris, where the oldest streets appear in dark red, notice that the vast majority of them still exist. Newer streets usually simply subdivided older areas (with the exception of Georges-Eugène Haussmann's major interventions that cut through several streets to reorganize the city).

to retract perpendicularly over a narrow strip that runs along the crack, the tension persists in a parallel direction. If a new crack nears an existing crack and enters the partially relaxed strip, then it suddenly changes direction, so that it comes to rest there by encountering the preexisting crack at a right angle. Indeed, cracks propagate perpendicularly to the tension that gives rise to them. Fissuring and reconnecting thus occur sequentially and are at the source of a universal property: cracks intersect at approximately right angles, irrespective of the nature of the materials or the fracturing mechanism.

Surprising as it may seem, such patterns of craquelures are not unique to materials that crack. Look at the map of a self-organized city such as Paris, in areas where zealous planners such as Baron Haussmann have not intervened (fig. 3). From a historical point of view, the network of streets that came into being as agricultural areas were divided, initially on the outskirts of the city, into smaller and smaller lots. These divisions are conveniently created at right angles to the boundaries of divided parcels, since it is a simple way to divide polygonal areas. Those dividing lines then naturally turned into new streets.

The urban network is thus organized sequentially, but also irreversibly: secondary roads can be connected to older arteries, but diverting an existing road is very rare. In fact, one would have to modify one or more already occupied, cultivated, or even built parcels. The map of a city's urban street network, if it was not designed at the time of a city's creation, will look very much like the craquelures in a painting.

EXPERIMENT

You can easily produce networks of cracks in your kitchen. For this, melt a mixture of a small glassful of sugar in five tablespoons of water in a microwave oven until it turns brown (1). Pour the mixture into a flat-bottomed container (2) and then cool it down by pouring cold water over it (3). That layer will instantly crack (4). You can also hear a crackle with each new crack.

A second experiment involves creating multiple cracks in a layer of corn flour mixed with water, which you've let dry on a plate. Adjust the proportions of water and corn flour to make the mixture pasty. If the layer is thin (on the order of 1 millimeter), you will notice a network of hierarchical cracks whose pattern is similar to the cracks you'd see on earthenware or an old painting. However, with a thick layer—and a bit of patience—you will see a network of small periodic columns perpendicular to the surface (below), which evokes the well-known basaltic columns of the Giant's Causeway in Northern Ireland. But that's another story!

4 | *This 1 centimeter-thick layer of dried corn flour reveals an astonishing fracture pattern. Periodic columns are formed perpendicularly to the surface.*

Materials

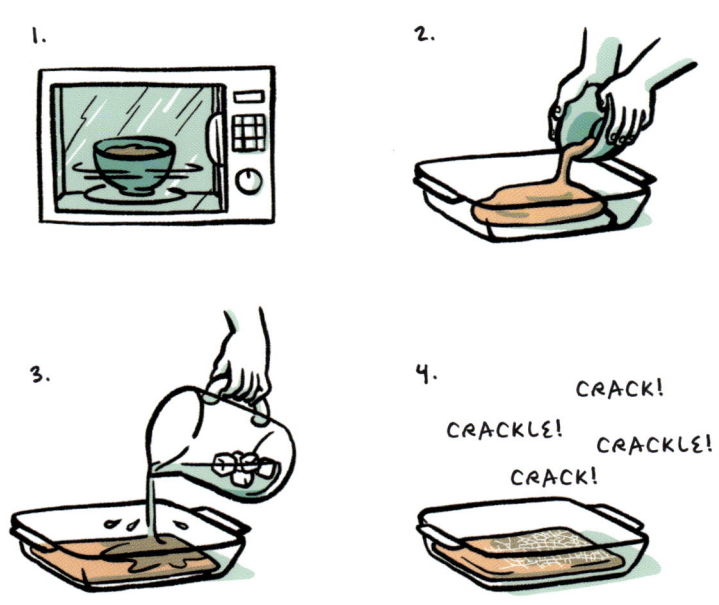

1.

2.

3.

4.

CRACK!
CRACKLE! CRACKLE!
CRACK!

WALKING ON EGGS

Why is it so difficult to break an egg by squeezing it in one hand? The geometry of the arc of the shell makes the egg-shape extremely durable. Can we meet the challenge of walking on eggshells?

What a terrible little sound,
the cracking of a hard-boiled egg on a tin counter;
what a terrible noise
when it moves in the memory of a man who is hungry.
    —Jacques Prévert, "A Plentiful Morning" (1946)

Walking on eggshells. This popular expression about emotional uncertainty hides literal possibilities. Before Easter it is customary to decorate eggs spectacularly using a variety of techniques. And that's an opportunity to concretely experience an eggshell's resistance. In Greece, as in

> 1   *Walking on eggs: is that literally possible? That's what this poster of an interactive exhibition suggests. The end-of-chapter experiment demonstrates that yes, it is!*

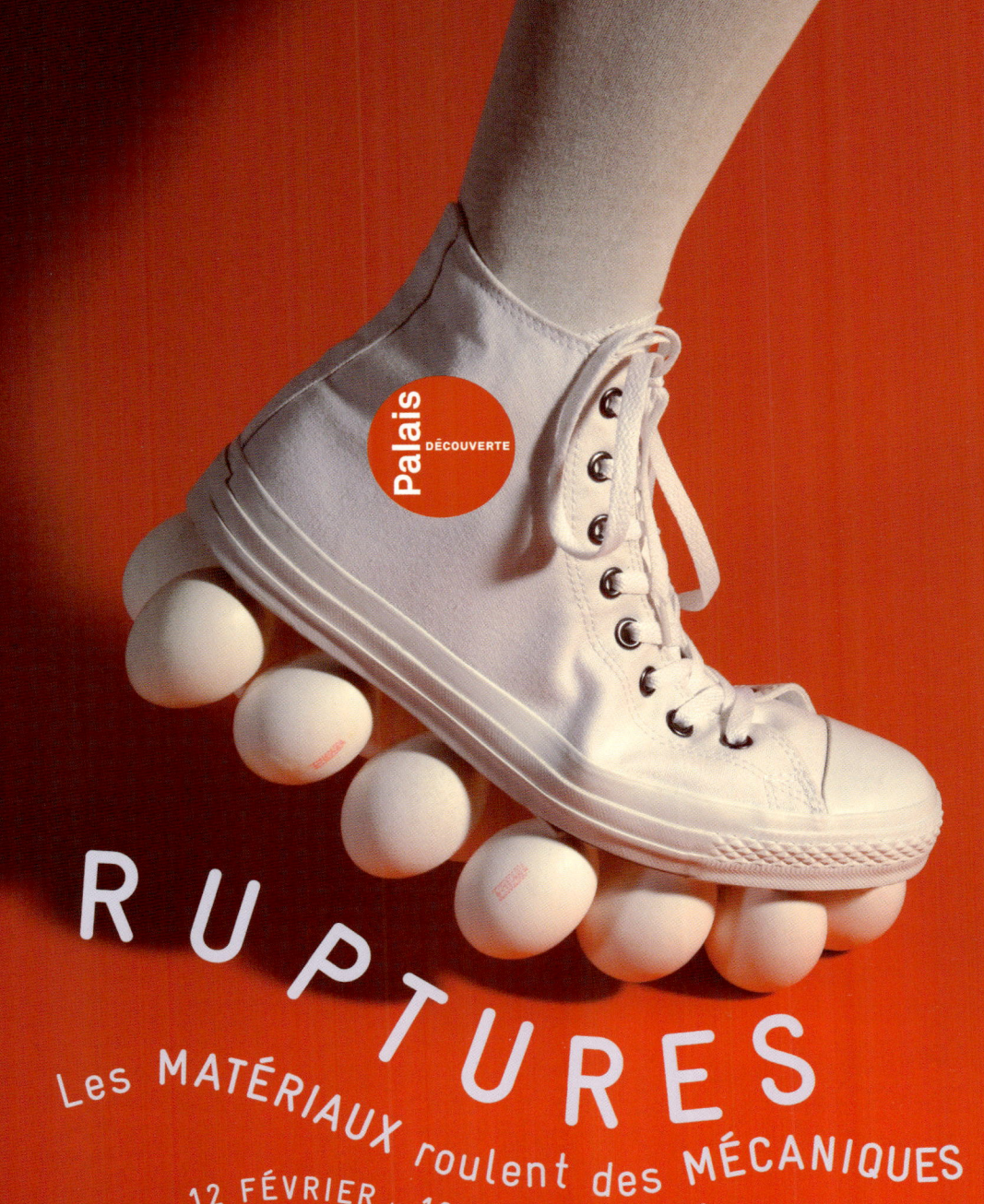

Palais DÉCOUVERTE

RUPTURES

Les MATÉRIAUX roulent des MÉCANIQUES

du 12 FÉVRIER au 10 NOVEMBRE 2013

other countries with Orthodox religious traditions, there's a fierce tradition that involves an egg battle. The competition takes place at a family meal, using hard-boiled eggs. Each participant knocks an egg against a neighbor's egg and, as usually only one of the two shells breaks, the one whose egg has remained intact is declared the winner. If you have a belligerent soul, know that it is more difficult to break an egg at its tip than on the rounded part. In other words, the side of an egg turns out to be the easiest part to break. What is an eggshell's resistance to deformation all about? The answer lies in part in its curvature.

Vault Effects

When you put pressure on an egg, its curved shape—its arched, vault-like shape—tends to redistribute the pressure throughout the shell. In architecture, this principle is used to build arches able to support heavy loads [ELEGANT STONE ARCHES]. It is thus very difficult to deform an egg by simply squeezing it in your hand along its axis. Similarly, in the kitchen, you always break an egg by banging it not only on the side, but against the edge of a bowl (or with a spoon for soft-boiled eggs) to concentrate the force on a small area and create a high, local stress that causes the shell to break.

To understand the strength of eggshells, MIT researchers have made semi-ellipsoids out of elastomers of the same base diameter and thickness, but with different curvatures at the top (fig. 2). You can make similar hollow shells by depositing a big drop of melted chocolate on a solid, egg-shaped surface: the drop spreads regularly over the contour of the surface and, when it cools, it ends up forming a solid shell of uniform thickness.

By pressing on the top of their strange, elastic shells, these researchers found that the most curved ones were also the stiffest. The explanation lies in the three-dimensional nature of the shell. Spherical shells are curved in all directions, unlike cylinders, and pressing them forces these curved surfaces to approach a plane. Now, as cartographers know, it is impossible to flatten the globe to make a map of it without distorting the

An egg with elastomeric shells of decreasing curvature of the same thickness and diameter at the base. The pointier hull is more resistant to caving in than the flattened hull.

continents [FOLDERS AND TAILORS: MASTERS OF VOLUME]. So, flattening a locally spherical shell doesn't just modify its curvature. It also forces it to stretch in its own plane. This is done at the cost of an elastic energy that grows as the shell's curvature increases. While all thin objects are easy to bend, they are much more difficult to stretch. And that's why it's easier to break an egg on its side than at its tip.

Imagine for a moment that these experiments might be reproduced with cylinder-shaped eggs. The change of curvature induced by the deformation would not necessarily require stretching and would be relatively easy. Conclusion: cylindrical eggs would be much easier to break!

Inverted Shells

Can we understand what happens when we put pressure on an eggshell using the elastomer model under increasing pressure? No, because unlike the real eggs whose shell is fragile, it can deform rather than break. That's the same as the case of hollow balls, such as those used in ping-pong games. At impact with a paddle, a ping-pong ball flattens before returning to its spherical shape after it is hit. But if it flattens too much, or if you inadvertently step on the ball, the flattening may indeed be irreversible. It then bears a concave shape that is practically symmetrical to the corresponding initial spherical portion (fig. 3).

Most of the deformation in the material is localized in the fold that is created around the inverted area. It is sometimes visible thanks to the

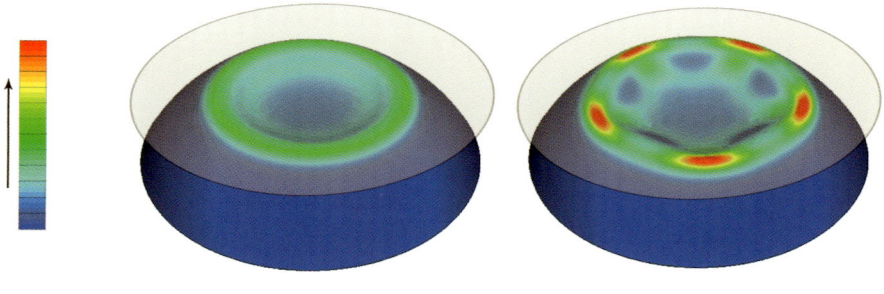

3   *Numerical simulation of the shape of a ping-pong ball that has been partially crushed by a flat plate. The color map shows increasing stresses from blue to red. Very quickly, we notice an inverted bump (left) taking shape. This small area is surrounded by an almost circular fold. Stronger crushers make it polygonal (right).*

white mark left on the damaged areas. You can try to salvage the damaged ball by immersing it into hot water. By heating the trapped air, you are indeed exerting internal pressure that is sometimes enough to return it to its original shape, although there will usually remain traces of the accident.

This situation where there is contact between a spherical cap and a plane is common, and physicists have studied it. Alas! It does not help us understand how an egg breaks when it is strongly compressed. In the Greek Orthodox Easter egg game, when participants hit their neighbor's egg with their own egg, is it always the egg with the more curved shell that best withstands the hit? At first it penetrates less than the opponent's does. But in this early phase of contact, the internal forces are actually similar regardless of the curvature. It is therefore the most fragile material, regardless of its shape, which ought first to give way. It is possible, however, that the break will only occur during subsequent stages of the crushing. This scientific problem remains unsolved.

Thus, although we understand what constrains deforming curved elastomer shells, predicting the outcome of a hard-boiled egg fight is far less obvious!

EXPERIMENT

An egg shell manifests remarkable resistance to breaking—and that's a good thing because otherwise eggs would break as soon as hens laid them. To test this resistance, get two egg cartons with six eggs in each, and make sure the eggs have no defects. Your challenge: to walk on the eggs without breaking them!

To accomplish this feat, gently place your feet directly on the eggs or on a light plate placed on top of the eggs. We also recommend wrapping the eggs in plastic to avoid damage in the event of any accident—a tiny crack or poor positioning could be tragic! The record to break exceeds 100 kilograms on a single egg.

Materials

GLASS
TEARDROPS

Elastic, hard, and brittle: glass presents properties that do not always seem compatible and yield unpleasant surprises.

The starry structure on a pane of laminated glass is often synonymous with vandalism, which does not preclude finding a sort of beauty in its unpredictable and disordered fractures (fig. 1). This was not lost on Marcel Duchamp when the glass of his famous *Bride Stripped Bare by Her Bachelors, Even* was accidentally broken: he immediately glorified this unexpected, accidental creation by renaming the work *The Large Glass*.

| 1 | *A laminated window after the impact of a 4-kilogram bullet arriving at the speed of 4 meters per second. The many fragments attest to the stresses within the glass. The intermediate sheet of the "laminated" glass prevents the scattering of fragments.* |

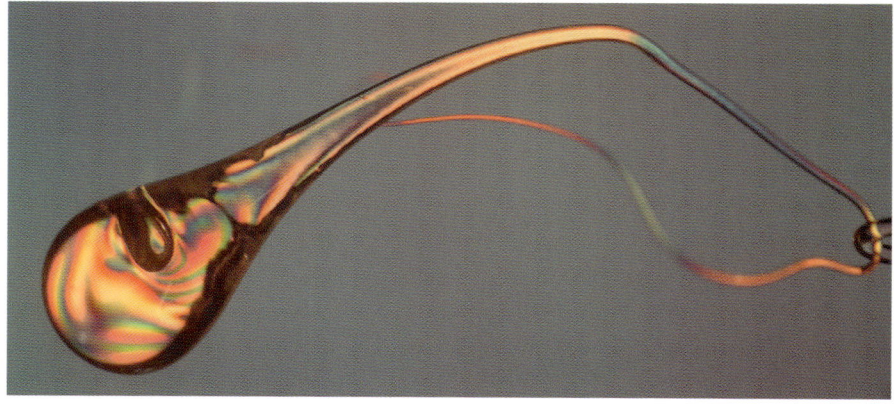

2 | Using a blowtorch to heat a glass rod will produce a liquid drop that flows as it stretches. Unlike a drop of water, which would very quickly become round, this viscous drop retains a teardrop shape. It solidifies as soon as it is immersed in water. Using crossed polarizers reveals magnificent colors—the signature of the internal stresses of the drop.

It is doubtless the supposed fragility of glass that strikes us the most, as glass teardrops perfectly illustrate (fig. 2). These impressive, elongated drops ending in a long, thin tail, not to be confused with marbles [STATES OF GLASS], seem to be frozen forever. They are produced by allowing a string of molten glass to flow into a container of cold water. Because of their original form, their transparency, and their ultimate colors, glass teardrops have long been fascinating. These drops also have an amazing property that has led to many industrial applications: gracefully slender as they are, they can still withstand the blow of a hammer!

The Achilles Heel of Glass Teardrops

Robert Hooke [NECKLACES AND CATENARIES] is credited with the first description of the mechanical properties of glass teardrops, which were then called "Batavian tears" or "Prince Rupert's drops." What is even more surprising is that although extremely strong, carefully breaking the tip of the thin tail of these teardrops (like a pharmaceutical ampoule) will cause a fracture to spread in less than one ten-thousandth of a second through the whole teardrop, which is soon reduced to glass powder. Why does it break this way? The secret of this strange behavior lies in the sudden cooling that the melted teardrop undergoes when entering the water. That has the effect of generating internal stresses between the skin and the heart of the tear. The slender tail is its Achilles heel: by breaking it, the tensile stresses present at the core of the material are released through fracturing. Multiple fractures propagate until the teardrop totally breaks into dust.

This spectacular experiment may seem anecdotal, but it relies on mechanisms that are now widely exploited by the glass industry to produce particularly tough, *tempered glass* objects. When the molten glass has been molded to endow it with its final shape as an object, it is first brought to a temperature of about 700 degrees Celsius, close to the temperature at which it starts to soften, before being suddenly cooled by jets of air. The surface of the material quickly solidifies, forming a more solid and elastic crust than its still-liquid core. As it in turn goes from a pasty state at 700 degrees Celsius to solid glass at room temperature, the interior of the glass shrinks a few percentage points and causes the already solidified crust to contract. The entire glass then becomes the seat of strong internal stresses: the surface is compressed, and the central zone is under tension. This ensures that the total balance of these forces across the thickness is zero.

How can internal stresses make tempered glass more resistant? Intuitively, one would expect stresses to weaken the glass. To understand this, then, we must take the example of breaking a typical glass plate. Such a break is caused by the propagation of a crack from an initial defect, most often a scratch on the surface of the object: a glassmaker thus begins cutting by scarring the surface using the tip of a diamond. In contrast, at rest the surface of tempered glass is compressed, which tends to close the precursors of cracking. To overcome this compression, one must therefore exert an effort. The inside of the material is, however, under tension; all it wants is to break if a crack manages to slip into its heart. But because its compressed shell protects it from external aggressions, tempered glass can be handled quite roughly before it ever breaks.

Ill-Tempered Glass

Widespread in cafeterias, tempered glass kitchenware is less elegant than the dishes that must have decorated Prince Rupert's table (Prince Rupert was a seventeenth-century German-English army officer, scientist, and colonialist who introduced England to a toughened glassware that was created by dripping molten glass into cold water). The undisputed success of this everyday glass is related to how it withstands impacts, thanks to a tempering process similar to that of Batavian tears.

We have all experienced this: when tempered glass tableware falls on the ground, it bounces a few times and survives the fall. But if it falls from a little too high, it will crack into many small fragments—the sign of the sudden release of the stresses stored in the material during tempering. This is another advantage of tempered glass: the many small chips reduce the risk of cutting oneself. This is a double-edged advantage, if you will, because it becomes very tricky to cut into or pierce such glass. Indeed, trying to do so usually causes microcracks that are likely to spread spontaneously, in a fraction of a second, reducing the glass to crumbs.

Communicating Glass

How do you make a window even more unbreakable? By imitating mother-of-pearl, for example [SHELLS AND MILLE-FEUILLES]. Bullet-proof panes thus consist of glass sheets separated by polymer films. The advantage of polymer is twofold: it reduces the propagation of the fracture to the thickness of the laminated pane, and it maintains cohesion among the fragments of glass. Applications for such laminated glass are not limited to vehicles' windshields. Without knowing it, you probably have laminated glass in your pocket: the touch screens of mobile phones are made of laminated glass and are often put to the test. When they break, these precious screens display the same starry pattern as a vandalized shop window.

Light is channeled and guided along optical fibers like a multi-colored water jet (below), while dimmed very little. At first a very popular decorative element, optical fibers have had a much more important—yet invisible—application: transmitting electronic messages over great distances. The huge rolls of cables that one sees along roadways under construction contain a set of optical fibers composed of a very pure and flawless glass. This glass is sheathed to protect the fibers and also to prevent light from leaking. It is this modulated light that carries information.

These fibers retain their mechanical and optical properties even when they are manipulated. Rodents are their biggest problem. But if you simply tie a knot in such fiber, and pull it, it will break very easily. Prince Rupert is never far away!

EXPERIMENT

Glass work is done at about 1000 degrees Celsius, so making glass tear-drops is not within everyone's reach. Instead, you can try making caramel teardrops. Heat sugar with a little bit of water in a saucepan (1). When the caramel starts to brown, pour a thin stream of it into a container of ice water (2). You will thus form caramel Batavian tears, which you will immediately remove from the water (3). In trying to break them, you will see that they give way suddenly and tend to shatter into myriad small chips. Enjoy the benefit of doing this with caramel instead of glass; these tear drops are sweet and edible!

Materials

1.

2.

3.

309

ADHERENCE
Force opposing the relative sliding of two parallel surfaces and characterized by a coefficient of friction.

ADHESION
Force opposing the separation of two surfaces.

AVERAGE CURVATURE
It is the arithmetic mean of the principal curvatures at a point of a surface. It is equal to 2/R for the points on the surface of a sphere with radius R, and is zero for soap films if the pressures are identical on both sides of the film (as in the case of catenoids).

BRITTLE BEHAVIOR
Describes the type of rupture of a solid that breaks abruptly without irreversible deformation: the resulting pieces match perfectly.

BUCKLING
Transverse deformation of a plate or beam, usually following longitudinal compression (beyond a threshold).

CAPILLARITY
Reveals the surface tension (or interfacial) forces present between two material phases (solid, liquid, or gaseous). It is responsible for wetting effects and comes into play in bubbles and foams.

CERAMIC
Terracotta object whose cohesion is obtained by heating.

CLAY
Mineral corresponding to a microscopic fraction (particles smaller than one micron) and that constitutes the main part of the soil's sedimentary rocks.

COEFFICIENT OF FRICTION
For a solid object sliding on a fixed plane, it measures the ratio between the normal force on the plane and the limit force which resists sliding (or *static* friction). When the upper block begins to slide, one defines a *dynamic* coefficient of friction, which is lower than the static case.

COHESION
Reflects the attracting forces exerted between molecules or, at a large scale, between particles of a solid or a liquid.

CONCRETE
A mixture of granular matter (and fibers) bound by cement and water.

CURVATURE
Any curved line drawn on a surface can be adjusted locally by an arc; the inverse of its radius characterizes the curvature of the line at this point. This curvature can either be concave (open upward) or convex (open downward). A full circle thus is said to have a constant curvature whereas a straight line has a null curvature. The curvature of the trajectory of a roller coaster alternates. This notion can also apply to a surface. In that case, two orthogonal "principal" arcs characterize the local topography. Their curvatures are in the same direction in the case of a bump and are in the opposite direction for a pass.

DEVELOPABLE SURFACE
Surface that can be made flat without stretching, tearing, or wrinkling it. Such is the case of a

cylinder or a cone, for example.

DIFFUSION
Microscopic phenomenon of transporting of matter or heat in which the flux of the transported quantity is proportional to the spatial variation of this quantity.

DROP and **BUBBLE**
A drop is a volume of liquid immersed in another fluid (air or a liquid with which it does not mix) maintained by its interfacial tension. A bubble is the opposite: a gas in a liquid. A "soap bubble" is like a drop in which a bubble is trapped. Such a construction is stabilized by the presence of surface-active molecules on both sides of the soap film.

DUCTILE BEHAVIOR
Describes the rupture phenomenon under a load that deforms a material irreversibly before it breaks.

DUCTILITY
Reflects the irreversible elongation of a material, which can lead to ductile failure. An archetype of a ductile material is chewing gum.

ELASTICITY (modulus of)
Coefficient of proportionality, characteristic of a material, between the stress applied to a material and its deformation, in the case of small deformations.

FLEXION
Transverse deformation of a rod or plate following a stress applied transversally, thus generating a modification of its curvature (without threshold).

FABRIC
Describes a thin, easily deformable layer made of intertwined threads.

FOAM
The result of aggregating a set of bubbles. In a dry foam, there is an extremely thin interfacial film, and the bubbles form a geometrically well-defined network.

FRACTAL
Characterizes geometries that are on average similar when studied at different magnifications. The coasts of Brittany, like the corrugated coast of southern Maine, are often cited as an example: their shapes are statistically identical, whether you consider one or tens of hundreds of kilometers.

FRACTURE (BRITTLE, DUCTILE)
Loss of cohesion between constituents of a solid, whether sudden or progressive or not (F) or not (D).

FRAGILITY
Aptitude of a body to break under sufficient stress without generating irreversible and often dramatic plastic deformations. This is typically the case of glass.

GAUSSIAN CURVATURE
This is the product of the principal curvatures (maximum and minimum ones) at one point of a surface: it is zero for a plane, a cone, or a cylinder.

GEL
In a network of polymeric chains forming links among themselves, the behavior evolves as the percentage of links

increases from a viscous regime (sol) to the state of a solid (gel).

GLASS
Noncrystalline ("amorphous") solid that characteristically transforms smoothly and continuously from a liquid state at high temperature to a cold solid state, called vitreous, without there being a clear transition between these two phases, unlike a usual sharp solid/liquid transition.

GRAIN
Isolated, solid, material element, large enough for thermal motion effects to be inoperative (typically bigger than 10 micrometers).

HARDNESS
Resistance to local mechanical indentation in the surface of a solid. Hard materials (e.g., diamond) scratch softer materials (e.g., aluminum).

HYDROPHILIC
Characterizes a substance or chemical group that has an affinity for water. This is for example the case of very clean glass.

HYDROPHOBIC
Opposite of hydrophilic (the nonstick coating of a pan for example).

INTERNAL STRESS
Reflects internal forces in balance within a solid and that manifest when the solid breaks (e.g., tennis racket strings).

LUBRICATION
Characterizes the relative slip between two solids between which is present a liquid or a gas, and which is a function of its viscosity.

MENISCUS
Curved interface between a liquid and another immiscible fluid.

NETWORK
Describes a set of links connected at nodes.

PLASTICITY
Describes situations where deformation increases irreversibly with stress.

POLYMER
High molecular-weight molecule formed by a sequence of small molecules (monomers).

POROSITY
Fraction of the volume of a porous medium filled with voids. Its complement to unity is its packing fraction or compacity, which is the fraction of the volume filled with material.

PRESSURE
This is the ratio of the force per unit area applied perpendicularly to the surface of a material: it corresponds to normal stress. Atmospheric pressure corresponds to the equivalent of a mass of 1 kilogram concentrated on a surface of 1 square centimeter.

PROTEINS
Macromolecules composed of amino acid derivatives present and active in living systems.

SANDSTONE
Sedimentary rock where grains of sand (silica) are stuck together.

SHEAR
Flow by which two layers of solid or fluid material slide relative to each other. The resistance in shear depends on viscosity for a fluid and elastic modulus for a solid.

SLENDER (GEOMETRY)
Geometry characterized by the existence of a dimension (a sheet) or two dimensions (threads) that are small with respect to the overall size of an object.

SINTERING
Applies to a solid material obtained by making grains stick together by heating or pressurizing them.

SOAP (SURFACTANT)
Short rod-like molecule consisting of a hydrophilic head and a hydrophobic, aliphatic (which likes fats) tail. Such an amphiphilic compound tends to become adsorbed at the interfaces and thus to decrease the surface tension.

SOLUTION
Liquid containing chemical species in dissolved form.

SOFT MATTER
An imprecise concept that encompasses matter with intermediate rheological behaviors between an elastic solid and a liquid.

STRAIN
The relative displacements of elements within a material, independent of global translations or rotations.

STRESS
Expression of force per unit area exerted on an interface perpendicularly to the surface (normal constraint), or tangentially (shear constraint).

SURFACE (INTERFACIAL) **TENSION**
Force per unit length exerted along the surface of a liquid (or along an interface). It also relates to surface energy per unit area.

SUSPENSION
Liquid containing solid particles. One speaks of EMULSION for drops of another liquid with which it is not miscible.

TENSION (Mechanics)
Force exerted tangentially to a surface; or, the opposite of COMPRESSION, tension exerts stress by stretching a material.

THERMAL MOTION
Spontaneous movements of small particles, which manifest a thermodynamic temperature T = t + 273 ° where t is the usual temperature expressed in degrees Celsius.

TRIBOLOGY
Science of interacting surfaces in relative motion.

VISCOSITY
Characteristic of a liquid or a gas that quantifies its resistance to flow at low velocities or small geometric scales.

WETTING
Ability of a liquid to spread over a solid.

YOUNG-LAPLACE LAW
Crossing an interface between two fluids or between a fluid and a solid results in a jump in pressure, the product of the interfacial tension γ by the average curvature of the surface. The pressure inside a drop with radius R is thus shifted by 2γ/R relative to the atmospheric pressure (and 4γ/R for a soap bubble because a double interface is then in play). In contrast, when the pressure is the same on both sides of a soap-free film, it has zero total curvature.

READING LIST

BUILDERS

> Bonner, J. T. *Why Size Matters: From Bacteria to Blue Whales*. Princeton University Press, 2011.
> Gordon, J. E. *The New Science of Strong Materials: Or Why You Don't Fall through the Floor*. Princeton University Press, 2018.
> Gordon, J. E. *The Science of Structures and Materials*. Scientific American Library, Times Book, 1988.
> Gordon, J. E. *Structures: Or Why Things Don't Fall Down*. Da Capo Press, 2003.
> Kakalios, J. *The Physics of Superheroes: More Heroes! More Villatopins! More Science! Spectacular Second Edition*. Avery, 2009.
> Lemoine, B. *The Eiffel Tower: The Three-Hundred Meter Tower*. Taschen, 2008.
> McMahon, T., and J. T. Bonner. *On Size and Life*. Scientific American Books, 1983.
> Thompson, D. *On Growth and Form*. Dover Publications, 1992.
> Timoshenko, S. P. *History of Strength of Materials*. Dover Publications, 1983.

CREATING SHAPES

> Baer, E., A. Hiltner, and R. J. Morgan. "Biological and Synthetic Hierarchical Composites." *Physics Today* 45, no. 10 (1992): 60.
> Ball, P. *Patterns in Nature: Why the Natural World Looks the Way It Does*. University of Chicago Press, 2016.
> de Gennes, P. G., F. Brochard-Wyard, and D. Quéré. *Capillarity and Wetting Phenomena: Drops, Bubbles, Pearls, Waves*. Springer, 2004.
> Edge, D. *The Curious Life of Robert Hooke: The Man Who Measured London*. Harper, 2004.
> Liger-Belair, G. *Uncorked: The Science of Champagne*. Princeton University Press, 2013.
> Ramirez, A. *The Alchemy of Us, How Humans and Matter Transformed One Another*. MIT Press, 2020.
> Stevens, P. S. *Patterns in Nature*. Little, Brown and Company, 1974.

BUILDING WITH THREAD

> Brunetta, L., and C. L. Craig. *Spider Silk: Evolution and 400 Million Years of Spinning, Waiting, Snagging, and Mating*. Yale University Press, 2010.
> Goodfellow, P. *Avian Architecture: How Birds Design, Engineer, and Build*. Princeton University Press, 2011.
> Treloar, L. R. G. "Physics of Textiles." *Physics Today* 30, no. 12 (1977): 23.

FROM SAND TO GLASS

> Challoner, J. *The Atom, A Visual Tour*. MIT Press, 2018.
> Guyon, E., J.-Y. Delenne, and F. Radjai. *Built on Sand: The Science of Granular Materials*. MIT Press, 2020.
> Van Damme, H. "Concrete Material Science: Past, Present, and Future Innovations." *Cement and Concrete Research* 112 (2018): 5–24.

MATTER IN MOTION

> Intaglita, C. "What Makes Sand Dunes Sing?" *Scientific American*, November 11, 2015.
> Lisa, M. *The Physics of Sports*. McGraw-Hill Higher Education, 2015.
> Sohn, E. "How a Venus Flytrap Snaps Shut." *Science News for Students*, January 28, 2005.

FRACTURES

> Eberhart, M. *Why Things Break: Understanding the World By the Way It Comes Apart*. Three Rivers Press, 2003.
> Goehring, L., and S. W. Morris. "Cracking Mud, Freezing Dirt, and Breaking Rocks." *Physics Today* 67, no. 11 (2014): 39.
> O'Brien, M. J., B. Buchanan, and M. I. Eren, eds. *Convergent Evolution in Stone-Tool Technology*. MIT Press, 2018.

ACKNOWLEDGMENTS

This book, in its variety, has benefited from the suggestions and contributions of many friends, academics, industrialists. We thank them all:

Romain Anger, Arnaud Antkowiak, Basile Audoly, Étienne Barthel, Pierre Bideau, Didier Bouvard, Marie-Ange Bueno, Philippe Claudin, Jérôme Crassous, Jean-Claude Daniel, Michel Darche, Régis Debray, Lucie Domino, Antonin Eddi, Vincent Floderer, Laetitia Fontaine and Aurélie Vissac (amàco), Yoël Forterre, Daniel Fruman, Pamela Goldin, Gustavo Gutiérrez, Xavier Guyon, Jérôme Hoepffner, Antoine Humeau, Véronique Lazarus, Gérard Lognon, Joël Marthelot, Xavier Müller, Xavier Noblin, Pascal Oudet, Guillaume Paoletti, Ludovic Pauchard, Jacques Pélegrin, Pedro Reis, Frédéric Restagno, Mathilde Reyssat, Romain Ricciotti, Benjamin Thira, Henri Vandamme, Jacques Villeglé, Jeremy Maxwell Wintreberg, Jean-Michel Wierniezky.

Some colleagues have generously provided illustrations for this book. The list of images below recognizes their contributions. The comics accompanying our experiments bear the touch of Naïs Coq.

Our colleagues at the PMMH (Physics and Mechanics of Heterogenous Matter) laboratory and from the national group MePhy (Mechanics and Physics) have encouraged us in our four-year excursion, with the myriad obstacles and steep slopes that would lead us to some picturesque looks at natural science, which we are glad to share with our readers.

We are grateful to the translation's editor, Jermey Matthews, and his team at the MIT Press that our book, originally published at Éditions Flammarion with the active collaboration of editor and physicist Christian Counillon, is now available in English. We are delighted to thank Patricia "Patsy" Baudoin for her fine translation of a work that is a little out of the ordinary in science-related literature, as well as Mike Woolf and Marie Yvonne Guyon for their immediate help in our exchanges about this English version of our book.